AF323429

THE
Design-Inspired
INNOVATION
WORKBOOK

BENGT-ARNE VEDIN

Royal Institute of Technology, Sweden

THE
Design~Inspired
INNOVATION
WORKBOOK

World Scientific

NEW JERSEY · LONDON · SINGAPORE · BEIJING · SHANGHAI · HONG KONG · TAIPEI · CHENNAI

Published by

World Scientific Publishing Co. Pte. Ltd.

5 Toh Tuck Link, Singapore 596224

USA office: 27 Warren Street, Suite 401-402, Hackensack, NJ 07601

UK office: 57 Shelton Street, Covent Garden, London WC2H 9HE

British Library Cataloguing-in-Publication Data
A catalogue record for this book is available from the British Library.

THE DESIGN-INSPIRED INNOVATION WORKBOOK

ISBN-13 978-981-4289-63-4
ISBN-10 981-4289-63-9

Typeset by Stallion Press
Email: enquiries@stallionpress.com

Printed in Singapore by Mainland Press Pte Ltd.

CONTENTS

Chapter 1

PROLOGUE: INSTANCES OF DESIGN — CASE STORIES

There is a world of fascinating designed objects, services, experiences, and emotions, and a host of wonderfully illustrated books on the subject. This chapter presents only a random few, with the idea that they serve as a backdrop for the remainder of the book. So they should function as the props of a scenery to be revealed, a reasoning to come — function as triggers for reflection, and stimulants to further ideas. And objections!

To be continued... the Vigix story

In *Design-Inspired Innovation*, Vigix founder and book co-author, Eduardo Alvarez, told the story about his business idea, and the associated product. The idea was to develop a system for the rental of DVDs that would require no human involvement; everything in the transaction would be handled automatically by a machine. To achieve maximum convenience, much-frequented locations would be preferable, thus where space is at a premium. Ergo: the machine must have a small footprint and be inexpensive, and also easy, fast, and inexpensive to re-stock. Together with design consultancy IDEO, Alvarez developed a scenario for how a family en route on the highway to some distant destination was halting at a Vigix kiosk to rent a DVD to keep the kids happy, the DVD to be returned by mail later.

To achieve user friendliness, Alvarez has a principle — a concept must pass a critical test group of one: his distinctly un-nerdy mother in Mexico. Yes, the concept did indeed pass that critical "My Mom's Test"!

The problem of high reliability for DVD delivery from the machine led to a breakthrough, a key invention, resulting in a patent application. The method, the technology for letting the customer receive the DVD, turned out to be a generic one, not applicable just for DVDs. Remarkably, it is a system without any moving parts — reliability to a fault — and much of the whole system revolves around this breakthrough.

So what happened to Vigix? Where is it today?

Alive and kicking, thank you. The concept has rounded out with sophisticated software for controlling a network of Vigix kiosks — and as so often with new ideas, Vigix is developing along a different trajectory from the one initially foreseen. Distributing DVDs may not be such a big market, with broadband Internet transmission as one of the alternatives. Instead, the generic technique for dispensing a product automatically from a mechanism, maybe a kiosk, opens up a host of opportunities, the correlation being that kiosks may be loaded and re-loaded by ordinary deliverymen from the likes of UPS or Federal Express. Among such opportunities, the Vigix home page lists prepaid mobile phones and accessories, which may be generalized to other electronic gears such as iPods; print media such as maps, books, and magazines; event tickets;

Figure 1.1.

greeting cards; and more. As we can see from Fig. 1.1, the Vigix kiosk is designed to occupy about the same space as a person.

MIT and personal transportation

At the annual Buckminster Fuller design competition, the winners in 2009 were a group from MIT, who presented an entire system for personal transportation, Mobility-on-Demand. The concept holds that there should be a number of docking stations (racks) for the storage, retrieval, and charging of the electric vehicles that constitute the system hardware. There are three types of vehicles, envisaged to cater to different transportation ranges — a scooter, a bike (possibly electric), and a minicab (Fig. 1.2). Their availability is monitored and their utilization managed by a computer system connected to those docking stations, also allowing users to reserve a vehicle from a net-book or a cell phone (Fig. 1.3). This computer management includes charging and acquiring payment for usage.

As can be seen in Fig. 1.4, the scooters, with electric motors fully integrated into their wheels, as well as the minicabs, are collapsible, making for compact storage. Safety, convenience, and comfort figure prominently — and fun as well. We can happily read in the presentation

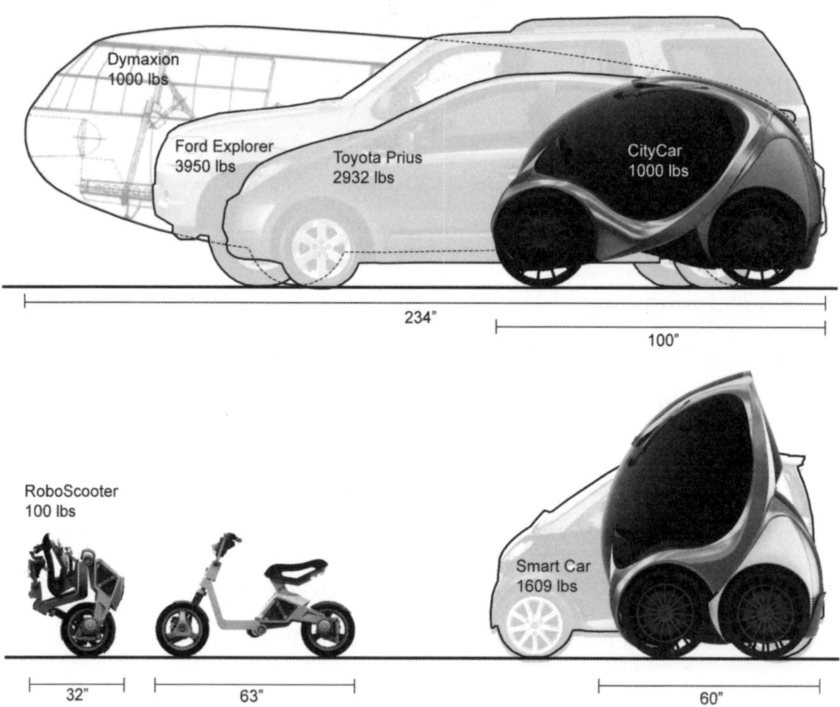

Figure 1.2. CityCar and RoboScooter Footprint comparison

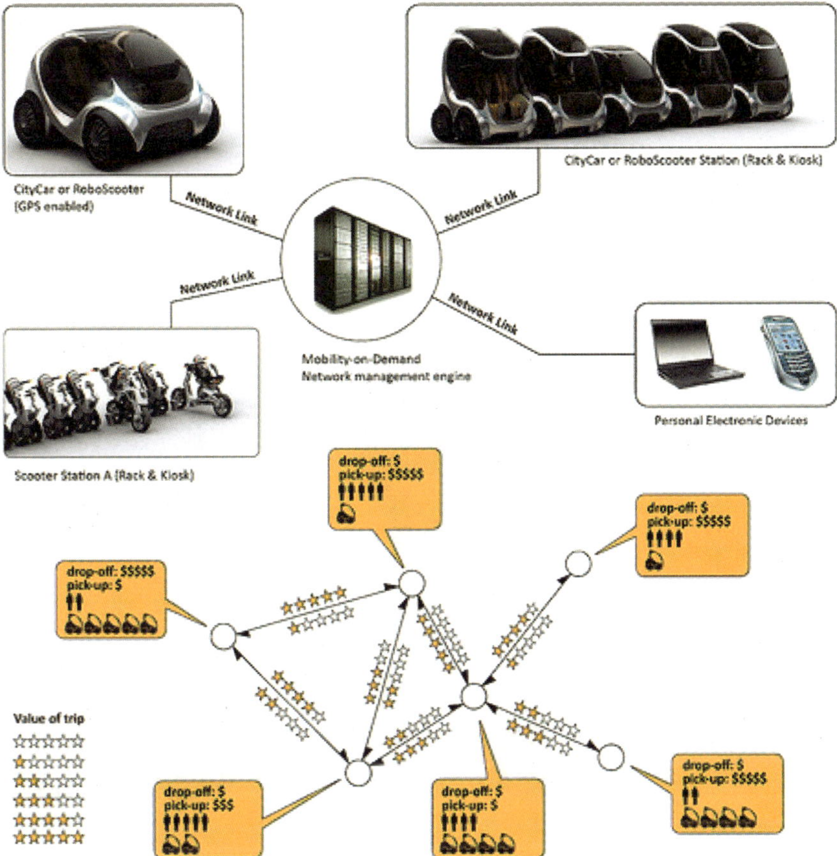

Figure 1.3.   Availability and utilization monitored by a computer system

"In-wheel motors offer an example of *design-inspired innovation*" (emphasis added). Those CityCar fully integrated in-wheel electric motors and suspension systems eliminate all need for traditional drive trains like engine blocks, gearboxes, and differentials; braking and steering is likewise included. The wheels can be re-directed omni-directionally (that is, 360°), allowing for extreme maneuverability.

Figure 1.4. CityCar Chassis — Full Scale Working Prototype

Merry furniture

The Spanish design firm Merry seems to specialize in absolutely intriguing, playful designs (appropriate company name, no?). Here is a dish inspiring you to take a (another) bite (Fig. 1.5); a zip lamp, where you control how much light to use by zipping up and down (Fig. 1.6); and you must admit that the little end tables (Fig. 1.7) are absolutely charming, because they are anthropomorphic, no? — not, for a Swede, the ordinary IKEA frugality.

Figure 1.5.

Figure 1.6.

Figure 1.7.

Nike+ into the shoe

Nike engineers, possibly inspired by the fact that in marathon races, runners are equipped with chips that help time them, discussed and brainstormed future intelligent running shoes. Runners often listen to music while jogging, relying upon an MP3-player like the iPod. Perhaps an iPod should be integrated into the runner's shoe?

One of the Nike designers had worked for Apple for a long time, so a contact between the two companies, and the relevant teams within them, came naturally. Together, they concocted storyboards telling how intelligent shoes might run, where one metaphor was 'the shoe speedometer,' recalling that initially, cars had no speedometers but now, of course, a car without one would be unthinkable.

Nike had come up with a sensor, but Apple took over that responsibility, miniaturizing and perfecting this component. Nike focused on the shoes and also on the interfaces to the iPod and the Web. Thus, the end result was a *system*, creating a log for the jogger as well as allowing her to hook up to an Internet community. In August 2008, Nike organized "the Human Race" in 25 cities around the world. There, runners could participate, individually, everywhere and not just in those cities, running 10 km and uploading their data to Nike. All in all, almost 800,000 people took part in that run. Of course, the community is open to more or less ingenious initiatives, such as people challenging each other: "first to 100 miles," "don't miss doing three workouts a week," "beat *the average per km time* for your age group" (not *everyone* can do that)...

For an individual running, the Nike+ helps keeping track of distance, speed, and may also help establish the route covered with the aid of maps on the Internet. The gadget may intercept the jogger, prompting if she sags and also activate some exciting music if such might be called for. Already a long time ago, biomedical research had established that the time that a runner's foot is in contact with the ground is inversely proportional to the runner's speed, independent of stride or slope. This had been frowned upon as not terribly exact, but turned out to hold well within error margins sufficient for the Nike+ function, a research discovery fundamental to the gadget's design.

The Nike+ consists of just three parts — an accelerometer, a transmitter for communication, and the necessary battery (Fig. 1.8). It has to fit into a shoe so shoes must be prepared with a small compartment

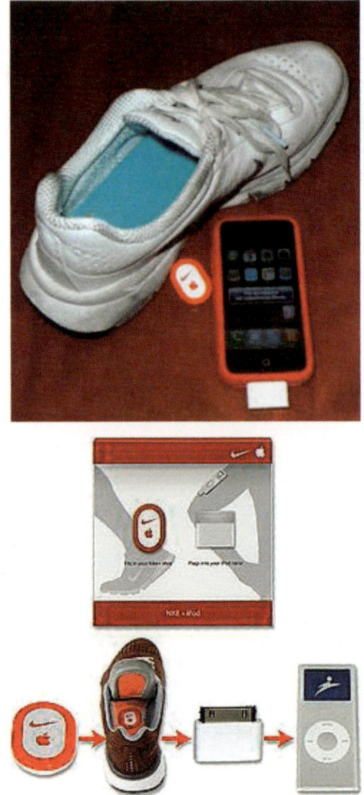

Figure 1.8.

for the Nike+ (Fig. 1.9). There is no heart rate monitoring, and no GPS tracking of the course run — that has to be done on the Nike web site. This equates with a lesson Nike learnt from Apple and the iPod: simplicity, focusing on user experience.

Nike has learnt a lot from users and their behavior. Not every runner has an iPod, so the alternative is a bracelet monitor (Fig. 1.10) where the data collection part is detached after the run and fit into the USB port of a personal computer. Data can then be uploaded to the runner's

Figure 1.9.

Figure 1.10.

personal file at the Nike+ site. On the site, the runner may follow her progress, establish goals, and compare with other runners or statistics of various types. One of the lessons for Nike was that people may try and test the site a few times but if they have logged on five times, then they are hooked and have become regulars. Or, perhaps, this reflects the positive habit-making effect that running — at least five times — has had on the individual?

As with the iPhone App Store, independent developers have established themselves, with open source initiatives such as Neki++ and freeware Running Tracker, making the user independent of the Nike+ site (Fig. 1.11). And for those who wish, Twiike can post their running data directly on to Twitter. Traditionally, Nike has been eager on patent protection but now seems to look rather positively on open source as a mechanism to enlarge the Nike community and thus the brand's appeal. Also traditionally, Nike focused on physiological needs — with Nike+ it is aiming at psychological and social demands, and, well, the *meaning* of running.

Shimano (re-)invents bicycle coasting

In *Design-Inspired Innovation*, we credited Shimano with having been instrumental in establishing "mountain bikes" as a market of its own, not by attempting to elbow itself into the place occupied by entrenched bike

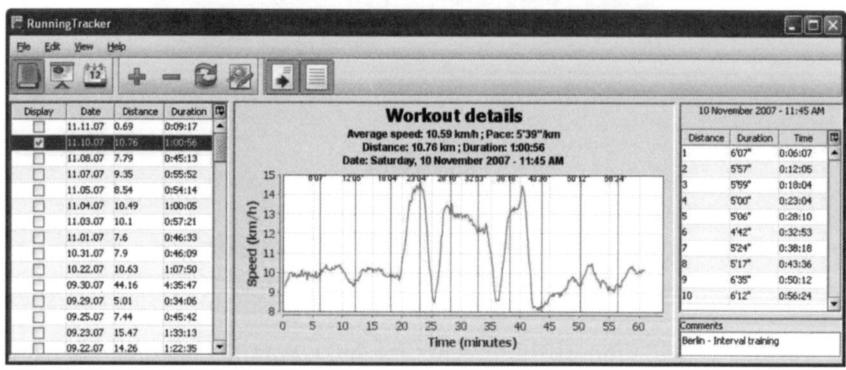

Analyze and **compare** your Nike+ workouts with Running Tracker !
 Import your existing workouts directly from the Nike+ website
 Import your workouts from your iPod Nano
 Analyze and compare your workouts
 Generate graphs and statistics

Figure 1.11.

manufacturers, but, instead, by establishing itself as the 'must-have' gearbox for mountain bikes. Studying and talking to bikers and ex-bikers, Shimano, together with design firm IDEO, discovered that the fast growing population of aging people held many ex-bikers and would-be bikers who felt gear-shifting going uphill necessary but too cumbersome. So, an idea was born: an automatic gear-shift for bikes, one that changed gears when the going got increasingly difficult (and shifted again, for the downhill ride).

An idea was born, and, like with mountain bikes, a new category: coasting. Together with IDEO, Shimano had developed the automatic gear-shift, and now set out to market the new approach to biking, and to convince the traditional producers of bikes to develop their own lines of Shimano-equipped coasting bikes. Three of the six largest bike manufacturers had introduced coasting bikes in 2007, and seven more producers did so in 2008.

The above was written before the appearance of IDEO CEO Tim Brown's 2009 book Change by Design.[1] Here, we learn that there are several more elements to the coasting bike design that evokes a 'traditional' and 'childhood' experience (Fig. 1.12). In addition, there is another important dimension: coasting bikes are sold through stores that are designed to be less off-putting to senior citizens than shops geared at extreme sports, with lycra-clad salesmen.

KOR re-designs water

In many parts of the world, water is a scarce resource — clean, portable water, that is. Often, it is consumed out of plastic bottles that are produced

Figure 1.12.

from petroleum in energy-demanding processes; even more energy is required for the transportation of these bottles. And out of those bottles, just one fifth is recycled, the remainder waste, often a nuisance or worse, such as, for example, with toxic stews amassing in the oceans.

KOR had a dream: to redesign water, or, rather, the ways water was bottled and consumed, and the way water consumers are drinking. The polycarbonate plastic in ordinary plastic bottles comes with traces of toxic substances. "Water ReDesigned" would have to be associated with design, health, and sustainability; the water bottles should be cool as well as reusable.

At the RKS design firm, KOR found the competency for discovering water users' unmet emotional yearnings and discover essential elements in the 'ritual of hydration,' of water intake. Available drinking/rehydration alternatives were plotted in a diagram with two axes, one being the degree of interactivity, the other being engagement or empowerment. Hydration/ drinking should tell a story, and KOR bottle users would feel like heroes relying upon a better, purer way to drink and to rehydrate. Ideally, they would function as evangelists starting a wave of viral demand (blog postings by bottle users indicate that the wave is on its way).

The eventual bottle should be something to be proud of, to feel good for both the one drinking water and in demonstrating her concern for the planet: a piece to display, to vaunt (Fig. 1.3). Water should be delivered easily and in one big gulp; the cap should allow for one-hand use and be impossible to lose; the bottle should be easy to refill under an ordinary tap, from a fridge, or a water cooler; and the material must be environmentally friendly and healthy.

While developing a guitar series, RKS had made good contacts with Eastman Chemicals, and so posed the question on whether they might propose some better alternatives to polycarbonate plastic. The answer was a qualified negative: 'not quite yet.' Soon, the Eastman Tritan™ co-polyester was introduced, featuring versatility, impact resistance, dishwater safety, wondrous clarity, furthermore allowing for varying wall thicknesses, and moldable in tools made for polycarbonate. Of course, it met the no-health-risk requirement.

It might sound trivial but the size of the mouthpiece, the opening which lets water into the mouth, is nothing but. Too large, and water will spill on the side; too small, and water will come intermittently if not sucked; the bottle squeezed or equipped with an extra air inlet, a troublingly cumbersome addition. RKS performed intensive

Figure 1.13. Photo credit: Ben Dowdy and Carla Olson of Eastman

testing to arrive at the ideal size and shape of the mouthpiece (Fig. 1.14). A side result was that ice cubes from the fridge now fit into the bottle.

Next challenge: the cap. To be operated by one hand, it could not be a screw-cork. After lots of experimentation in the RKS lab, the solution was a cap opened by the simple pressing of a lid by the thumb.

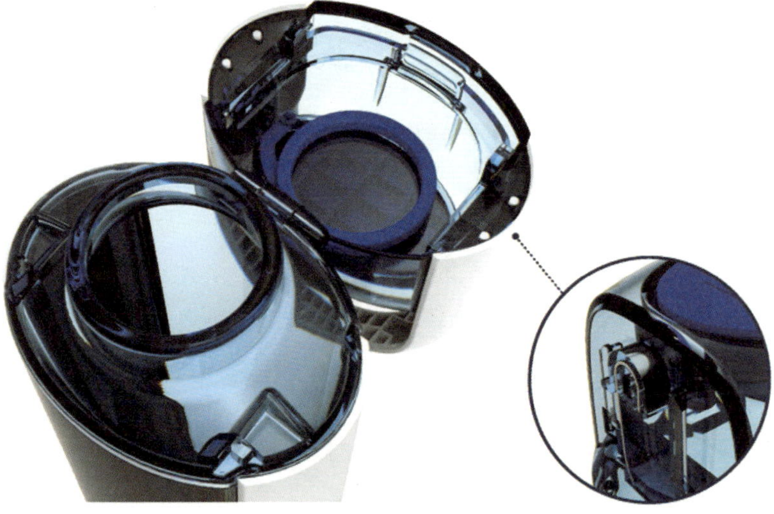

Figure 1.14.

A built-in gasket prevents any leaking whatever bottle position once the latch has been re-set and the bottle sealed again. The resulting top resembles a Chinese puzzle (Fig. 1.15).

Another partner was also involved: Nypro, experts in manufacturing "integrated plastics solutions." Since the designers wanted the bottle bottom free from the normal notch in the middle of the bottom, emanating from the plastics production process and the gate for the material, the gate was moved off-center, to the side. Thus, the bottle — the water — would look really clear, clean, and crystal-like.

Normally, re-usable water bottles would not offer much beauty or excitement. The KOR one, however, proposes a drinking, re-hydration experience, with a bottle material with attractive tactile properties: a glass clear material with but a tint of blue, and an obelisk shape that intrigues (there are three more colors available, each with its own non-profit organization

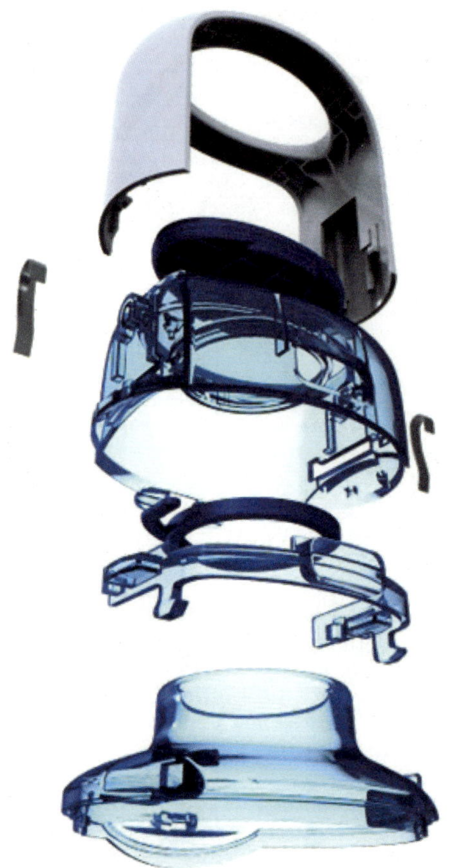

Figure 1.15.

as a partner, like the Container Recycling Institute for Sunrise Orange). The frame is a semi-transparent white aimed at evoking the feel of a glacier; functionally, the frame offers an easy grip for the user's finger(s) (Fig. 1.16).

What about personalizing your water bottle? The designers decided to include another unique feature: the KOR stone. This is a small round piece, a disc-shaped token, that can easily be inserted into the cap, by lifting its flexible end seal; a stone printed with content chosen by the

Figure 1.16.

user. The stone's text can be seen just by the person drinking the water. (Fig. 1.17) Any bottle comes with a set of stones with different messages and new ones can be produced and exchanged over the KOR web site and within the KOR community.

Philips re-thinks MRI

The Dutch electronics (and light-bulb) giant Philips has been a pioneer in applying industrial design, recently also with an explicit focus

Figure 1.17.

on simplicity, with a director in charge of precisely the latter. At a conference in Copenhagen, Hans Robertus of Philips Design made an instructive exposé of Philips design activities, including some enticing light-bulb designs, based on light-emitting diodes, with which we begin our report from Philips (Fig. 1.18). So my use of the term 'light-bulb' is, in fact, obsolete.

Following Hans Robertus, we present some slides from the design process for a magnetic resonance system, MRI — an environment for radiology: images of the body's interior. For magnetic resonance tomography to produce images akin to those resulting from X-rays, the body (or parts of it) has to pass through the appropriate apparatus. The term 'tomography' indicates that computer power is relied upon to compose an integrated picture from measurements that amount to 'one

Figure 1.18.

slice at a time.' The part of the body studied is on a conveyor belt passing through an opening whose walls contain the necessary magnetic resonance generation and image signals collection equipment.

Figure 1.19 shows a photo of the radiology system configuration as it existed, the sketch (Fig. 1.20) shows a vision of a putative future solution. The team had a truly eclectic composition: health-care

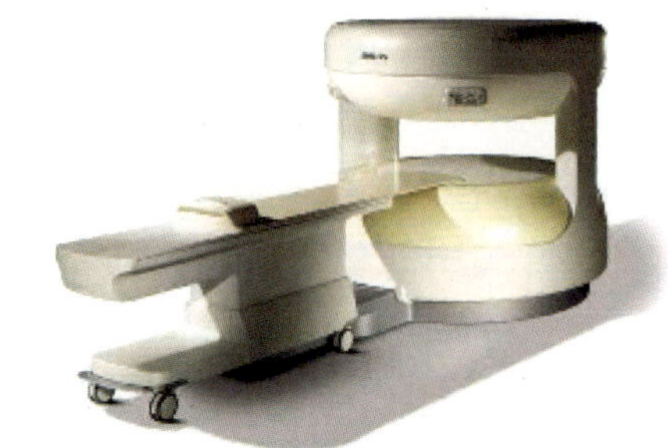

Figure 1.19.

Figure 1.20.

expert; psychologist; engineer; architect; product; interaction; multi-media; and light designers. The very activities envisioned were described in storyboards; as we see in Figs. 1.21–1.23, they are different for different potential stakeholders, and several categories of medical personnel and patients, including children. The different 'flows' applying to these categories were made to correspond — that is the idea — to specific designated areas and activity zones. One particular concern, not shown here, had to do with managing the coils for the magnets. Finally, there is a resulting concept specification where hand-drawn sketches have been substituted by CAD renderings (Fig. 1.24).

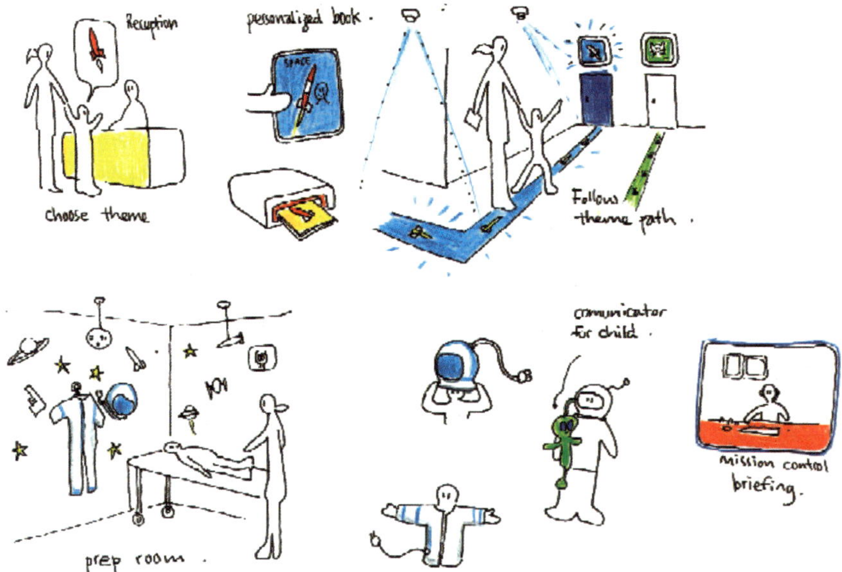

Figure 1.21.

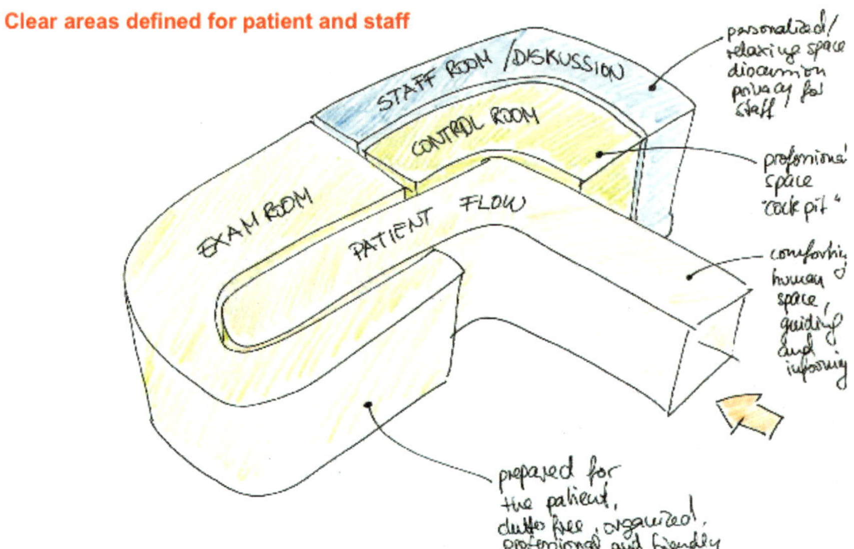

Figure 1.22.

Figure 1.23.

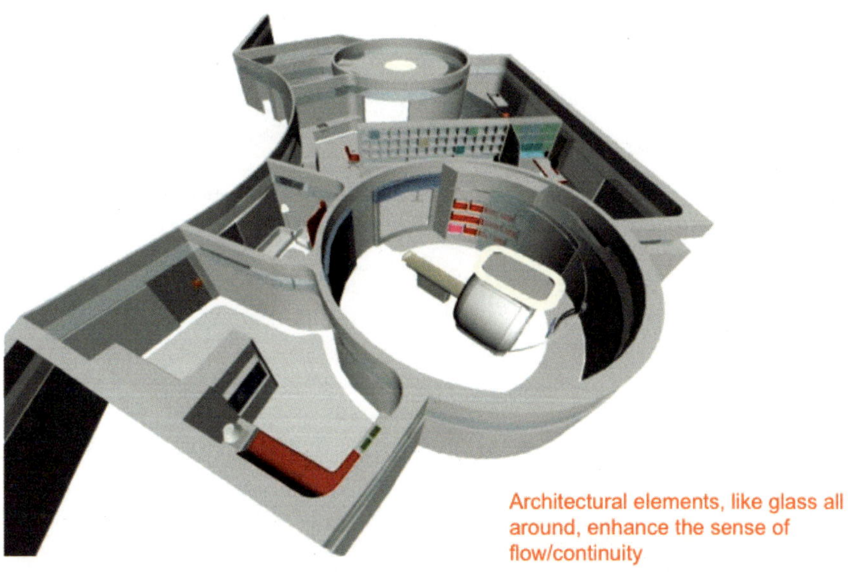

Figure 1.24.

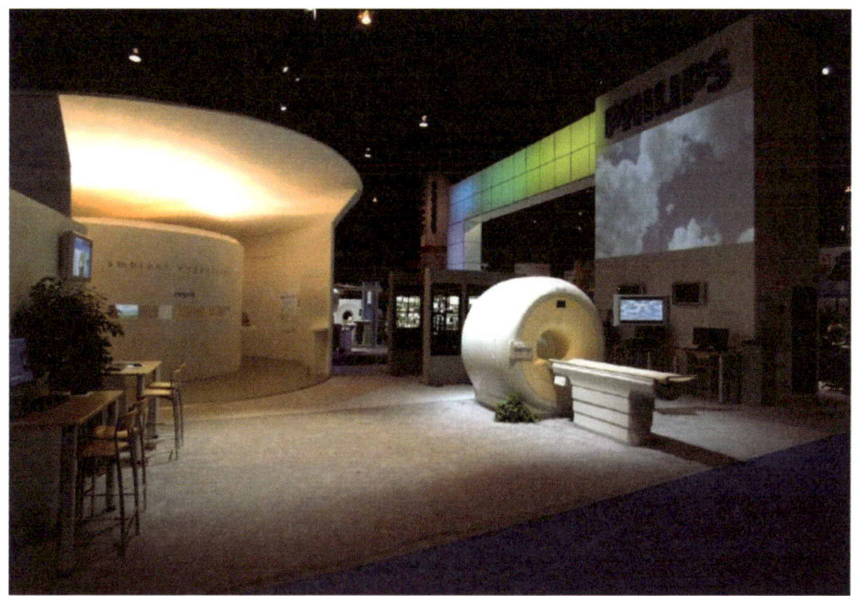

Figure 1.25.

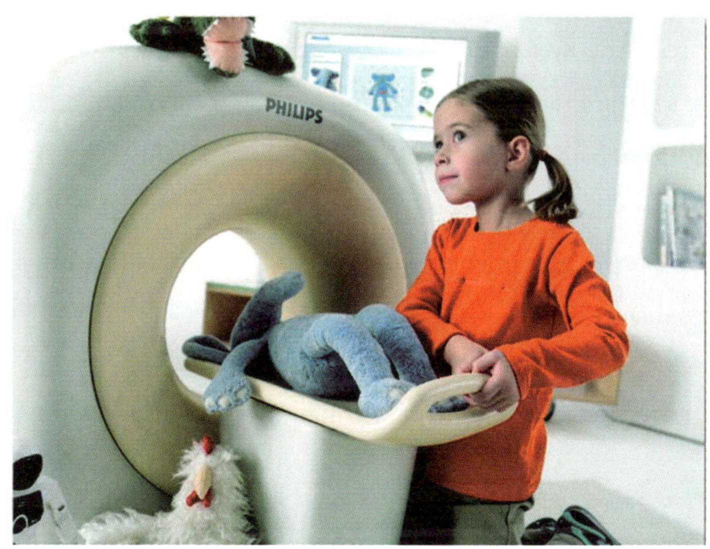

Figure 1.26.

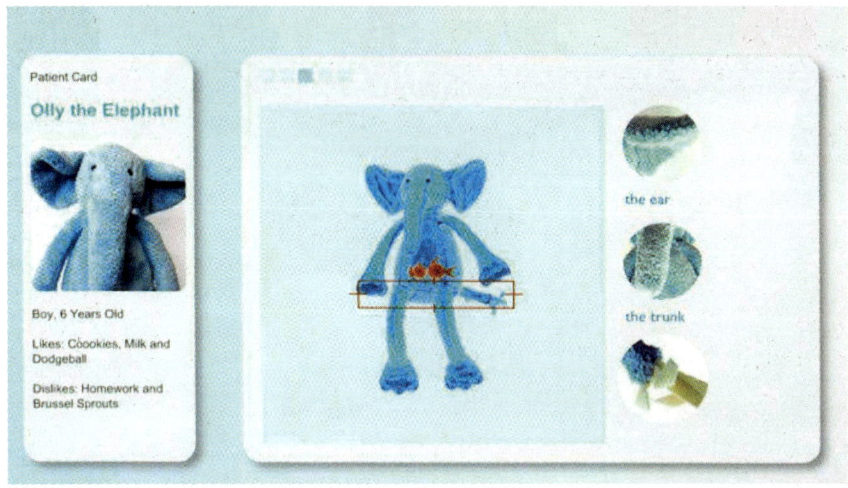

Figure 1.27.

Kids were among the groups consciously considered (Figs. 1.26–1.27). And whom would they care for if not their best friends? And sketching is a faculty that develops early in life...

The resulting system showed impressive advantages over its predecessors: sedation rates were down by 30–40 percent, radiation dosage was reduced by as much as 50–75 percent, capacity utilization showed an increase by 15–20 percent, images were of higher quality, and the steps in preparing a scan could be reduced, leading to higher patient and staff satisfaction.

Coca-Cola freestyle and design machine

Coca-Cola got a new design chief, David Butler, who first did not have that title, and who met a design ignorant company — so he wrote a three page memo, describing to his colleagues that Coca-Cola, as a matter of fact, is one of the world's largest design firms. This was the wake-up call for a troubled company; Butler envisaged that design could come to its rescue. Supply-chains had problems, bottlers made trouble, opportunities in global markets were lost, core products were losing traction. How to reduce carbon footprint from transporting sodas around? How to handle increasing consumer skepticism to carbonated sodas? How to attract consumers through variety, and how to know what they would prefer? With between 400 and 500 brands in more than 200 countries and more than 1.5 billion servings of the company's products every day, managing complexity is compulsory. The new emphasis on design resulted in two early initiatives with major potential. One project aims at supporting individual bottlers all over the globe, the other offers the consumer a soda delivery system.

To allow even the most minute marketer to shape customized point-of-sales material, Butler commissioned what is named the Design Machine, a Web-based tool (Fig. 1.28). This is modularized for scalability and allows for producing material tailored to local markets, still in line with global brand strategy. There are 36 languages to choose from, more than 100 product options, almost 9,000 templates, and also a variety of drinking occasions. The tailoring process takes a couple of minutes and shaves off 30 percent of fees to local ad shops. Market feedback for local tweaks is collected and analyzed automatically, generating improvements.

The other item, directed to consumers, is the Freestyle fountain machine (Fig. 1.29). It offers customers a very broad selection of drinks while eliminating the previous 5-gallon concentrate bags, replaced with thirty 46-ounce inkjet printer cartridges, an idea — micro-dosing,

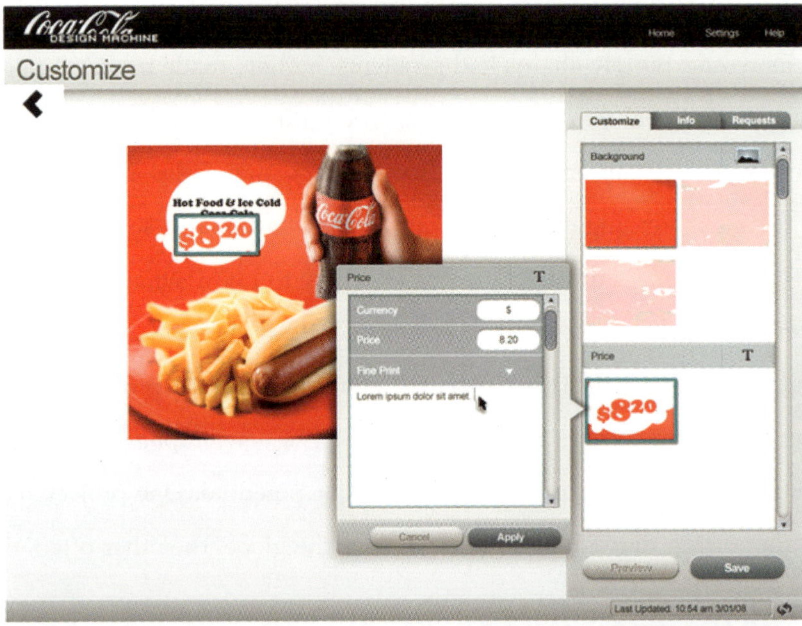

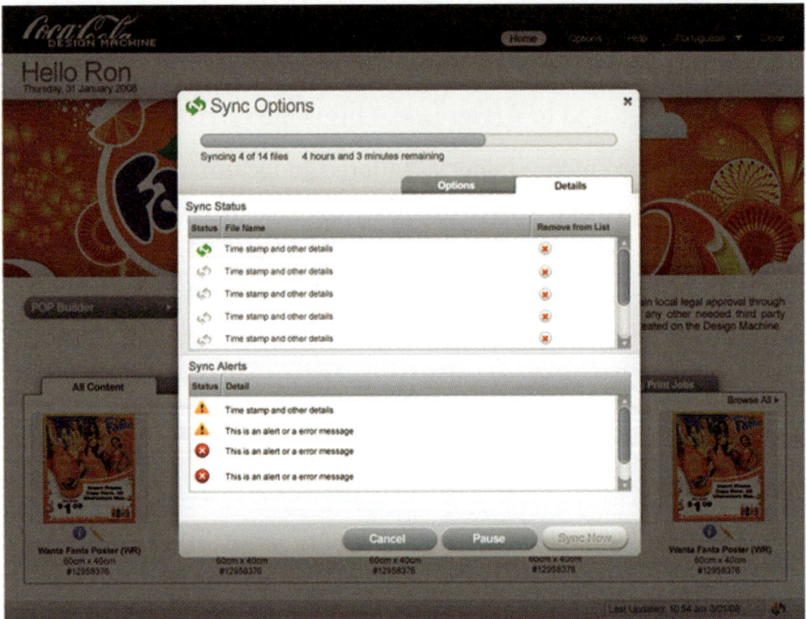

Figure 1.28. The Coca-Cola Design Machine user interface.

Figure 1.29.   The Coca-Cola Freestyle: most of the radical change is found in the interior, though there is a touchpad for choosing — composing? — one's drink

with just drops for flavor — borrowed from the medical industry, thus bringing down carbon footprint vastly. Cartridges, equipped with radio frequency ID chips, track instantaneously what is being dispensed and an integrated computer monitors consumption, enabling staff at the Atlanta HQ to analyze data about beverage popularity, peak times, and preferred locations. Company staff can also talk back to the machine, letting it know digitally if a particular flavor needs to be discontinued, recalled, or tweaked. Test marketing is easier and less expensive to carry out, and the performance of a new offering or campaign can be traced closely. The data collected allows distributors to fine-tune inventorying and ordering. An early discovery is that Diet Coke without caffeine has a sales hike in the late afternoon.

The Freestyle machine heralds a new business model: a delivery system, rather than a machine, has been designed. It is the same size as previous vending machines, those delivering soda cans of some ten different varieties. The consumer will now navigate to her choice by an easy-to-use touch-sensitive screen, offering 140 varieties from combinations of the drops from the thirty cartridges, so that, say, raspberry is used in a Coke, a tea, or flavored water (Fig. 1.30). The screen makes

Figure 1.30.

promotion and information messages possible and allows the consumer to inspect the product and its ingredients before purchase, selecting a drink for its brand, calories, caffeine content, and color. As a network node, the machine can communicate digitally not only within Coca-Cola but also, using Bluetooth, with the consumer.

Unusually for Coca-Cola, the Freestyle is an internally generated product, resulting in a number of patents. The company is even manufacturing the machine internally, though with hardware from suppliers such as Samsung.

The British Design Council offers a series of case stories for design-oriented companies, telling about how they organize for and manage design: http://www.designcouncil.org.uk/About-Design/managingdesign/Eleven-lessons/.

Endnotes

[1]Brown, Tim, 2009. *Change by Design*. Harvard Business School Press, Boston, MA.

Chapter 2

INTRODUCTION: WHAT THIS BOOK IS AND HOW TO APPROACH IT, WORK WITH IT

Design is a powerful force for the ignition of innovation. That was the thesis proposed and supported in our book, *Design-Inspired Innovation*. Design was seen as connecting more directly to the often unmet and unknown yearnings of the customer, the user, evoking in her delight, pleasure, even suggestive dreams. The support for our notion of design as an innovative power varied from case stories, insights from companies and designers consciously pursuing a route of design-driven and design-inspired innovation (DII, in brief), from the clustering of design firms around particular nuclei, the analysis of how components might contribute to product families, platforms, and brands, to how design could be seen as embodying particular values and meanings.

This book is a companion volume to the previous one, attempting to offer practical advice on how our findings might apply successfully: it is a Workbook. In order to walk the talk (odd metaphor, no?), it is organized to provoke the reader to be *inspired by design* — preferably by the book itself, engaged in experiencing some of the tenets of the lessons to be learnt and practiced. Therefore, introductory words to explain what you, the reader, will be invited to participate in: a journey.

Largely, the DII Workbook depends upon its predecessor, and thus its authors, Eduardo Alvarez, Sten Ekman, Susan Sanderson, Bruce Tether, Roberto Verganti, James M Utterback, and myself; more about this can be found in the Acknowledgements at the end of the book. Other sources of inspiration and facts are also found there; here, I want to get on with what I perceive as the reader's main purpose — to embark on the story. Please recall this book's characteristic of being a Workbook, not a piece of research, though it is based upon such, as documented in *Design-Inspired Innovation*. The Workbook, though, has been composed to be read without any consultation of the previous book.

As any object, the DDI Workbook is the product of design, in Bruce Tether's term, 'silent' or not, and, apart from the publishers' art directors' work on graphics and layout, mostly text or content design. A design, then, with the purpose to instill in you, the reader, engagement, enthusiasm, motivation, inspiration — and understanding. A design, hopefully, also to provoke, to invoke objections, since one of the messages is that creativity is stimulated by constructive criticism, and that success may come out of objections resulting in the identification of counter-trends or micro-segments, micro possibly turning macro, counter possibly turning mainstream.

It is suggested that the reader apply DII in her actual reading and operate on the book's lessons and advice. I underlined the fact that the book is designed, text and content subject to the designing. In our previous book, one central tenet, and a stark warning sign, emerged from recent discoveries in psychology, such as those pertaining to what is termed 'verbal overshadowing.' In brief, this means that if we have memories of something physical, like a human face, and if we then get a verbal description of it or are engaged in producing a verbal one, our original visual memory is erased, overwritten, and indeed overshadowed. This is obviously of importance for industrial designers sketching and brainstorming, but not just for them. So, one challenge offered to the reader would be: undertake to translate the book's words into vivid visual enactments (or not just visual), attempt to have the words recede, supplanted with perhaps dreamt-up tactile, olfactory, palatal, or musical sensations. Maybe, instead of a book, you could envisage a physical product, a service, a theatre piece, a melody, a fragrance? How would the book, or a chapter, or an argument, be perceived, or written, by a child, a poet, inspire a sculptor? If the argument, chapter, paragraph were a car, of which make; if a flower, which one; if an animal, or a sport?

I invited you, the reader, to partake in a journey, and tour through these pages — their content, their messages, their lessons, relying upon a well-worn metaphor. But then, when I said to be eager to 'get on with the story,' I had actually left the realm of metaphor and introduced an increasingly popular idea: the fact that we tell stories — tell stories when we establish brands, when we suggest catchphrases for companies and products, indeed, when we *design*. As we shall see, a dogged, thorough, and profound analysis of novels, films, and myths has reduced the basic plots employed to just seven, and a very condensed version of that analysis will be found in Chapter 6. There are quite a few novels that combine more than one of these seven schemas. So when I encourage and challenge the reader to discover to what extent elements of our book — the reader's and mine — coincide with one or more of these plots, interpretation becomes essential, as in design.

The reader would be eager to find a foolproof prescription for how to produce great products that were always guaranteed of success. We too! Needless to say, such recipes do not exist. Much is a question of timing, much a question of the competitive ecology when launching the new product. So this is not a book on innovation or innovation management, nor one on creativity, nor on design. There exist a plethora of such books and when we broach the theme of working creatively, say, or the one of design processes, the aim is not, and cannot be, to offer an exhaustive rendering.

Rather than thinking about success, as the ideal and impossible innovation management handbook would suggest, we should think about failure. To fail often, to fail inexpensively, to fail so that we learn a lot — to fail profitably, as it were. Evidently, this is not a recommendation to willfully create products that are sure to bomb in the market, but rather the springboard for establishing criteria for ventures worthwhile to launch, successful or not, but worthy of the benefit of the doubt. Nassim Nicholas Taleb in his great book, *The Black Swan*,[1] suggests that Black Swans — unusual, unforeseen events like successful breakthrough products or economic catastrophes — are, by their very nature, impossible to foresee or plan for. One should abstain from placing huge bets on either those perceived as "safe" ones, which may only have the potential of turning into negative Black Swans, or in sectors that carry huge costs from collateral damage caused by negative Black Swans, all the while making small bets on many potential Black Swans of which only the rare one will actually give any return, but then a huge one.

So those are our failures: small bets on potential Black Swans, with a learning experience thrown in for good measure. Do not overanalyze; it is more like firing before aiming, learning from what the firing gains.

Those shadows of verbal overshadowing vary. Lera Boroditsky at Stanford University and MIT has studied how language influences how we think. Here are just snippets but with a clear message: "treating chairs as masculine and beds as feminine in the grammar [turns out to] make Russian speakers think of chairs more like men and beds as more like women." Key is masculine in German, feminine in Spanish. Describing a key, German speakers were "more likely to use words like 'hard,' 'heavy,' 'jagged,' 'metal,' 'serrated,' and 'useful,' whereas Spanish speakers were more likely to say 'golden,' 'intricate,' 'little,' 'lovely,' 'shiny,' and 'tiny.'" And, mind you, all testing of German and Spanish speakers was performed in English, which lacks grammatical gender. The effect is at display in art galleries where Russian painters most likely paint death as a woman, and German painters as a man, reflecting the gender of death in the respective languages. This holds for as much as 85 percent of such personifications. Also, colors have been demonstrated to have different connotations in different languages.

So, to better understand this book, those findings induce me to tell just a little bit about myself. My mother tongue is Swedish, which, for example, features four grammatical genders; we dropped verbal plurals in the 1950s — but obviously those are just random facts, indicating that the reader sometimes may have to be wary more about what I have missed to elucidate than what is found in the text. By the same token, I offer the information that the country where I was brought up was indirectly but profoundly affected by WWII (we have had no wars since 1815), secular but with a strong Lutheran tradition. Sweden is one of the world's countries most heavily dependent upon international trade; it features the highest taxation in the world; and a strong social service. Abstaining from bifurcating into a handbook of Swedish statistics, let me just add that the female employment rate is often quoted as the largest in the world. Now, these scattered facts do not help you one bit; they only serve as a reminder that anybody's background is a subterranean world that we, at least, have to recognize the existence of. Successful design sometimes is successful precisely because it strikes home here, deep down.

And who are you, the reader? I should take the end user's view, should I not? As designers would tell, it is oftentimes productive to create some concrete representative(s) of those to whom the design is aimed. Such *personas* should be rather specific, as we shall see; here, again avoiding to preempt a later chapter by producing it prematurely, I will be but sketchy. One of you is Charles 'Chuck' Pearson, hailing perhaps from the US mid-West, serving briefly in the military before graduating with an MBA from… where? After a brief stint in the new business department of the X Company, you spent a couple of years in sales and sales supervision at Y Inc in S-town in up-state Y state. There you met your wife… kids… sports… moved on, VP of a small company; VP of a large company, as in all instances, a mechanical industry; now on a special assignment to a) introduce DII, and b) doing so by managing an important project. The other reader has a different background: she is Donna Blossom, black, from Virginia, a designer graduating from Z design school, then spending a few years in the European design hotspot Q; you are now about 35, running a large, quite famous industrial design firm fighting to assist large customer firms achieve DII. Again, personas should be more detailed. But would the book be different if the seasoned mechanical industry VP were a 55-year old black woman? Would its design?

This last question might belong in our chapter on applied creativity for DII; as mentioned, the detailed introduction of persona production will appear in a later chapter. It is in the nature of an introductory chapter that it serves as the hub for a hypertext web; there are links to any number of different more profound expositions. Such hyperlinks, advertised or not, will appear throughout the book; that simply follows from the nature of the subject.

The reader and the author also have to agree on what the book really consists of; what is the real product, the meaning inherent in the story, what does it evoke? Is it a proposal, competing for attention in a crowded attention space, offering an experience, or perhaps even calling forward a transformation? For Chuck, perhaps a transformation; for

Donna, a proposition. Every reader, however, must attempt to interpret and apply the lessons offered, the triggers prodding into concrete action, into undertakings and activities. That will call for prioritization, and it will call for creativity.

Since the workbook is generic, it is not the workbook particularly for web design, for the wheelchair builder, for the architect of furniture, or for the producer of new automobiles. Thus, again, it is up to the reader to adopt and adapt, to build upon — and to protest, since what is highly appropriate for one industry might be dangerous or counter-productive or just beside the point in another industry — market, user group, culture, nation, niche, demographic group. As we will highlight, constructive criticism may be utterly productive, and so might protests and objections. Again, creativity is called for, again, there might be an idea to find underlying principles that result in something substantially different in a situation vastly at odds from one described in this book.

The word 'workbook' implies that, ideally, the book should be read with pen, clay, Styrofoam plus knife in hand. Sketching is central to design; trying, experimenting, and doing are essential to learning, and to gain experiences. There is no need for the publisher to include such tools with the book since they are abundantly available (paper and pencil certainly are). Besides, they would differ from reader to reader, perhaps because of different access to resources but, first of all, because of those differences in 'location' in the space of niches, industries, cultures, individual preferences. Styrofoam or rapid prototyping, CAD/CAM or doodled sketches?

Another challenge has to do with where this book starts and ends. Its physical pages are well defined but the concrete examples and abstract discussions relate to its environment and its context, the reader's environment and context. There is no fixed boundary but rather a dialog, though the context must be said to be ambiguous, and, often enough, variable over time.

'It is up to the reader to adapt and adopt,' 'the book should be read' with a tool at hand, and then it is up to the reader to define the boundaries of the book! Is the author somehow abdicating, leaving the composing of the book to the reader, who, after all, paid for the product? Yes, I am abdicating! It is your book — a Workbook, by design.

To introduce the author in pictures, these represent some of the places where this book took shape — apart from plain offices… Watercolors by Gull-May Holst.

Endnotes

[1]Taleb, Nassim Nicholas, 2007. *The Black Swan*. Random House, NY, NY.

Chapter 3

COUNTING ON DESIGN: ARE THERE ANY PROFITS IN DESIGN-RELATED INNOVATION?

To give the store away directly: the response to the heading — yes, of course there are! Would this section be included otherwise? But the point is, rather, that this positive response emerges out of the very same concrete realities that underlie the entire book.

Possibly, the following short recap of figures is unnecessary. Because, as Roberto Verganti states so forcefully and beautifully,[1] figures can come only afterwards; for a design project, to ask for the financial value of a specific investment in design is futile.

Singular instances and examples

As such, and isolated, these prove nothing. The design-inspired change, developed with the assistance of consultant firm RKS, in a particular product (a washer-dryer from American producer Amana) cost all of 30 cents. And it allowed for a price increase by 25 percent or 100 dollars, reflecting boosted value in money terms in the perception of customers-users. Another corporate success that has been credited to design (but also to open innovation; thus, it may be the result of good management in general?) is the Procter & Gamble of then new CEO A G Lafley (more of him presently): while the S&P index shrank by 20 percent from the beginning of 2001 to 2006, P&G saw a value hike by 72 percent.

Design and global competitiveness

The 2009 International Design Scoreboard from Cambridge University, which explains the UK focus, provides a rather special relationship: twelve countries have been judged on a number of indicators and ranked as to their design capability which has then been plotted against relative global competitiveness in 2007, as per the World Economic Forum. As we can see (Fig. 3.1), it turns out to be a straight line, with the usual deviations from it by certain points, in the two-dimensional diagram, so better competitiveness correlates with higher — ranked — design capability.[2]

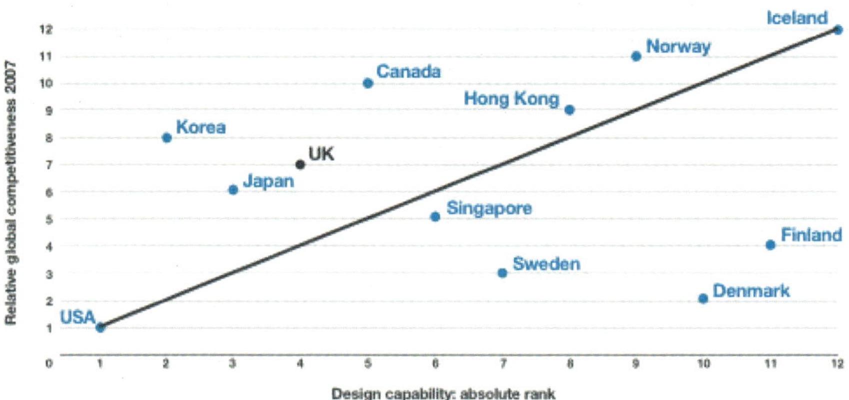

Figure 3.1.

Stock market performance: Design versus overall indexes

The statistics available are not up to date, but comparisons have been calculated between a stock value index comprising companies winning the annual Business Week IDEA competition and the ordinary indexes, S&P 500, Dow Jones, and NASDAQ. With 1994 as the base year, the design winners outperformed all three indexes 1994–1998 when these doubled — while the design winners' index tripled. During the bubble years of 1999–2000, NASDAQ outperformed the design winners temporarily. But when the bubble burst, NASDAQ sank back again to, or even slightly below, the level of Dow Jones and S&P, all three back down at their 1998 level in 2002, whilst the design winners now were at four times the initial value. In the latest year available for these statistics, 2005, the indexes showed a rebound to 2.7 of initial levels, while the design winner index had performed much better, having increased by a factor of six.

British statistics, compiled by the UK Design Council, tell a similar story: over the period 1995–2004, the general UK stock market index (FTSE 100) displayed a value growth of somewhat below 50 percent, while emerging companies and design dependent ones both had a value 3.5 times (that is, plus 250 percent) the initial one. Design-oriented companies outperform other firms in net income and earnings.

Firms' perceptions: Denmark

The Danish National Agency for Enterprise and Housing has commissioned a thorough study of the effects on corporate performance of employing design.

Danish companies acquiring external design competency recorded a 22 percent boost in gross revenue over five years. If also internal investments in design were accounted for, as much as a 40 percent *additional* revenue hike could be registered, compared to companies where design efforts were flat or reduced. Firms consciously committing themselves to design turned out to have 34 percent of their revenue generated from exports, as compared to 18 percent for those 'low' on design; this may, of course, to some extent reflect a structural difference. As expected, more revenue equates with more hires.

Firms were asked to tell how design was looked upon and treated, where, in a hierarchy, step 1 equated with no or inconspicuous design, step 2 with styling as a finishing touch, step 3 with design applied early on, and step 4 with something almost akin to design-inspired innovation, design here also pertaining to more than just one product, involving strategy and the corporation at large. Revenue was higher and exports significantly larger the higher on the design ladder a company had advanced. Firms with a strong commitment to design constituted a minority, though. Over time, there had been a migration towards higher steps on the ladder. Larger companies were more often found higher up, and 'high up' was also associated with higher export ratios, better economic performance in general, and hiring (the report does not discuss the direction of causal dependences — difficult to discern, certainly).

Firms' perceptions: The UK

British data tell of design stimulating exports: 51 percent of The Queen's Award for Export Achievement winners in 2002 directly attributed overseas sales success to investments in design. Over 90 percent found design to be valued and important among their international customers and fully 86 percent indicated that design was instrumental in making them competitive internationally. The study also revealed a higher design intensity among rapidly growing businesses (cf. further British data below), which were six times more likely than stagnant ones to see design as integral to their operations.

The UK Design Council has produced a 'Value of Design Factfinder' (2007). It provides detailed results on how UK companies perceive the value and impact of relying upon design. Almost half (46 percent) of all UK businesses believed that over the past decade, design had become more important in helping them maintain a competitive edge, and two thirds of UK businesses believed design to be integral to future economic performance. The vast majority (84 percent) of rapidly growing businesses agreed.

One salient fact is that for every £100 a design-alert business spends on design, its *turnover* increases by £225, and for every £100 a business corresponding to the 'design alert' moniker spends on design, its *profit* increases by £83. In businesses with design integral to operations, over three quarters claim to have increased their competitiveness and turnover through design, while they do not need to compete on price as much as others; less than half of these firms compete mainly on price, compared to two thirds of those who do not use design.

Almost a third (30 percent) of those design-alert businesses saw profits increase by more than the amount they spent on design in the very *same year*. Whether other design alert firms had held their position thanks to design is a moot question. For some businesses, profits rose as much as, or more than, their turnover. Here, design is directed towards trimming costs or holding market share rather than just boosting sales.

Almost half of UK businesses believe design to contribute to increased market share (46 percent) and turnover (44 percent), at least to some extent. Over the previous 12 months, many businesses attributed a substantial role to design in their achieving increased market share, competitiveness, and increased profits (percentages in the 40s).

Recognition of the contribution of design grows with the size of a business, from 44 percent among small businesses to 56 percent among medium-sized ones, and 77 percent among large businesses.

Adding value to products and services makes businesses more successful — and the more value they add to their core products and services, the better they do. But when they use design, specifically, to add value, they do even better in a whole range of ways, including bigger profits, revenue, and market share. 46 percent of those businesses planning to add more value in the future used design, the compared to just 6 percent for those that did not employ it.

Businesses with design integral to their operations are twice as likely to have developed new products and services. In the past three years (as seen when the study was performed), four fifths of them had done so, compared to a UK average of 40 percent. On average, design-alert manufacturing businesses develop as many as 25 new products a year through design. Over two thirds (72 percent) launch at least one, the median number cited was three so the distribution is decidedly slanted — as should be expected. Quite naturally, businesses with more employees and larger turnover tend to launch more products.

One of the results was that 80 percent of design-led firms had seen market share increases compared to 46 percent for the other companies; whereas the relative percentages for the opening up of new markets again were similar, 80 to 42 percent. Still on average, the fact that design-alert businesses open up new markets equate with an increase of their market share by 6.3 percent through design. Those in the retail, wholesale and leisure services, and in manufacturing saw the biggest increases, 6.9 percent and 7.2 percent respectively. While turnover growth is more likely to happen to businesses increasing their investment in design, the opposite is also true: those firms that decreased investment also cut their chances of growth.

In the services sectors, design-alert businesses develop, on average, two new services a year (larger firms, 2.6) through design though two thirds (68 percent) do not launch any at all. Businesses in the property, finance, and business services sectors are the ones most likely to introduce new services, averaging 2.2 a year.

Compared to others, a design-alert business is…

— … twice as likely to have developed new products or services recently; over the previous three years, nearly three quarters (71 percent) of design alert businesses had done so (UK average for all firms: 40 percent);

— … twice as convinced of design becoming more important to competitiveness; four out of five (79 percent) design alert businesses felt that design's importance to competitiveness had risen over the previous decade, while for all of the UK, only 46 percent of businesses believed this to be the case;

— … twice as likely to have increased its investment in design; over the previous three years, 63 percent of design alert businesses had invested more in design, something that held for just 31 percent of other businesses;

— … more likely to apply design to developing new products; twice as many design alert businesses did so, or 55 percent compared to 28 percent for businesses in general; they were also twice as likely to apply it to R&D: 21 compared to 12 percent;

— … more likely to use external design services: almost all (94 percent) design alert businesses had used design services in the previous 12 months, compared to two out of three (68 percent) of businesses in general;

— ... more likely to use, in particular:

- communications design: 68 percent for the 'design alerts' against 52 percent for others.

- digital and multimedia design: 45 percent against 31 percent.

- product and industrial design: 45 against just 17 percent.

Rapidly growing businesses were three times more likely than others to consider design crucial to success, and they were nearly six times more likely than static ones to see design as integral. Those firms rapidly growing were twice as likely as the UK average firm to have increased their investments in design, and more than two thirds had done so recently.

Over two thirds of manufacturing companies believed that it was well worthwhile investing in design in their sector. 79 percent of them (that is, their representatives) see design as essential for future performance, compared to 65 percent for UK businesses in general.

43 percent of businesses that used a designer to add value reported a great impact on competitiveness, compared to 25 percent of those that did add value but did not use a designer. Revenue turned out to be boosted by 40 percent when adding value relying upon a designer or designers, versus 20 percent when adding value otherwise. The corresponding figures for gaining market shares were 35 against 14 percent. Thus, 53 percent of businesses that used a designer added a lot of value to their core product or service. In businesses that did not use a designer, only 39 percent added 'a lot' of value to their core product or service. Furthermore, businesses that used a designer to add value were much

more likely to be planning more added-value activity in the future —
61 percent reported to be doing that, compared to just 34 percent of
those which did not use a designer. Businesses that employed a designer/
designers were even more likely to perform well when it came to adding
value to their core product or service. 69 percent of these businesses
had developed a new product or service in the previous three years,
compared to 47 percent of businesses that did not use a designer.[3]

Lots of figures and comparisons, telling of averages but not of
particularities. What might be the lessons for you, for your activities?
The development of relevant metrics? The collection of case stories to
learn from? The initiation of benchmarking efforts? Contacting design
consultants to learn about whether they have some hard, or soft, data
telling about the return on design investments — and of factors influencing those returns?

Do designers do design thinking?

When we published *Design Inspired-Innovation* in 2006, it was not exactly
out of a void; it was the fruition of several years' work. More than a
few design consultancies such as Design Continuum, IDEO, and frog
had highlighted the power of design in furthering innovation. Design
Continuum had coined the phrase the Design-Inspired Enterprise,[4]
Hargadon, amongst others, had delved into the links between design
and innovation,[5] several of our chapters leaned on our own previous
research and articles, and government backed agencies in, e.g., the UK
and Denmark had commissioned important studies, as demonstrated

above. And Hatchuel, and Barton, and Durling, and Sanders, and Pink, and Burnette, and... on and on. Owen had outlined his ideas about design thinking already in a conference keynote in 2005,[6] published in 2006 and 2007.[7] Buxton's 2007 book, *Sketching User Experiences*, has a title that not directly reveals that it features design thinking as an important theme.[8]

All that said, 2009 qualifies to be termed the year of design thinking[9] — at least until we see what the future brings. That very term, design thinking, is the explicit theme for at least three books, and several more evoke the same broad territory. Sensing its topicality, the Sloan Management Review compiled a Special Report on the concept:[10] "Design thinking — distinct from analytical thinking — has emerged as the premier organizational path not only to breakthrough innovation but, surprisingly, to high-performance collaboration, as well." Likewise for Matthew May, blogging on the subject August 3, 2009.[11] In a way, Warren Berger with *Glimmer*[12] is the most radical when that he suggests that we learn and adopt methods from designers to design even our lives, our social relations, our education (it builds partly on designer Bruce Mau's 2004 book *Massive Change*,[13] where he proposes "design of the world"; a new Mau book is advertised for 2010). Designers, Berger contends, are good at asking 'stupid' questions, and on thinking visually and laterally.

Thomas Lockwood is the editor of two 'design thinking' books, one co-edited with Thomas Walton, on *Building Design Strategy*[14] published in 2008, the other, *Design Thinking*[15] outright 2009; he is president of the

Design Management Institute (the DMI 2008 conference had Design Thinking as its theme). Hartmut Esslinger's *A Fine Line*[16] is yet another book in the same vein; A G Lafley's 2008 *The Game Changer*[17] is a practitioner's creed, and an important practitioner in this context, at that. Together with Lockwood, Tim Brown[18] of IDEO (and his colleagues, including the Stanford d.school), and Roger Martin, dean at the Rotman School of Management at the University of Toronto, are perhaps the most vociferous promoters of this currently hot topic.

Martin, in his book,[19] regards design thinking as a management style, not as about design *per se*. It is an approach, Martin contends, that strikes an even balance between analysis and intuition (designers would be too much intuitionists), and an approach that favors validity (that which works) over reliability (that which has been shown to work in the past; established templates). Instead of relying upon deduction, from known principles, or induction, from known facts, design thinking emphasizes abduction or the reliance upon educated guesses, ideas, or hypothesis that have to be tested. The fact that Martin sees designers as handicapped when it comes to 'design thinking' has made for questioning the choice of descriptor. Another discussion point has been whether this really is something new; in fact, it is not, but the spotlight and the discourse that it has attracted are.

Reliability is about

- repeatability which means predictability
- based upon past success
- thus the outcome is certain

- the number of variables is limited
- objectivity, little judgment
- quantity, that which can be measured

Validity is about

- validation from what happens in the future
- context dependency
- an array of many and diverse variables
- subjectivity, judgment
- uses, among others, analogy
- the qualitative

Currently, 'everyone's' hero as a CEO is A G Lafley (who retired by yearend 2009), the author quoted above and who, more importantly, was the one to turn Procter & Gamble into a growth hothouse instead of an also-ran (Martin has another hero firm also: Research In Motion, the Blackberry maker). Lafley's example demonstrates why adopting design thinking — for want of another term — is a challenge to organizations: slipping back into reliability would be all too easy and comfortable for, say, the economics department that is fairly isolated from the market. Claudia Kotchka, whom Lafley recruited to head P&G's design effort is a case in point: she nixed the job twice before accepting it, upon insisting that Lafley really back up the undertaking. And she is not a designer by training, she is an accountant.

Brown's of IDEO definition of design thinking seems to be evolving: in a Harvard Business Review article,[20] it is "a discipline that uses the designer's sensibility and methods to match people's needs with what is technologically feasible and what a viable business strategy can convert into customer value and market opportunity." Technology and business sense, talent from the broadest possible pool, early and frequent prototyping (though prototypes early on are more of mockups), and, not least, seeing with the eyes of the user are themes elucidated in Brown's book. One example of the eyes of the user is when a hospital patient gets a video camera attached to his head so that the design team can really see, and experience, exactly what s/he suffers.

As hinted, Brown has come to question his own definition as perhaps too constrained: "On reflection, this is a narrow description that focuses on design thinking's role within business. The next sentence that I wrote '...*design thinking converts need into demand*', which I borrowed from Peter Drucker, broadens things out a bit." Brown now leans to a broader area of application for design thinking than just business; the argument that it is better to leave the concept only loosely defined.[21] For Brown, in contrast to Martin, design thinking is a lot of what designers do (as for Durling, already in 2003).[22]

"T-shaped people" is another important dimension for Brown, that is, people who have some deep specialization, combined with broad general schooling. (Bill Buxton has topped it up by asking for I-shaped people, with their feet firmly in the mud, in practice.)[23]

One reason to look upon 'design thinking' as the wrong moniker is taking seriously Martin' contention that designers themselves are no design thinkers. The very fact that star example Kotchka of Procter

& Gamble is not a designer underlines the protestation. Another argument hooks on to the suggested T-shape (Kotchka!) and reformulates design thinking into something like, e.g., hybrid thinking, multidisciplinary capacity, boundary spanning, multi-versatility: combining different fields of thought.

To refer back to the number-filled first half of this chapter, during Lafley's regime (2000–2008), P&G revenue more than doubled from $40 B to $83 B, with earnings shooting up from $2.5 B to in excess of $12 B.

A book that strongly stresses the difference between *design-driven* and *designer-driven* is Verganti's *Design-Driven Innovation*,[24] with some of its punches but far from the whole powerful argument foretold in *Design-Inspired Innovation*. I once suggested that he should rather call it "meaning-driven innovation" though I understand why a publisher would not agree to that; the forceful argument is that those responsible for true innovation rely upon design to propose new meanings to customers. The meaning of a lamp is not that it is a lamp but that it brings light; the Kartel 'book worm' is not about books but a statement about its owner; the Nintendo Wii engages the whole body of the computer game player, not just eyes and thumbs.[25]

To arrive at those proposals, a design-conscious firm has to rely upon a network of interpreters: people working in design, but also trend analysts, journalists, fashion and show people. This should be part of an ongoing design discourse that can be both open in the sense of partaking in various activities happening in society and proprietary such as by organizing seminars and workshops, inviting to prototyping and suggestions, see box on Philips Design probes. This discourse has no well-defined boundaries and one can argue — Verganti certainly does — that there is the need for an inventorying of relational assets such networks, structures, or discourse substrates. Experimentation and research are important as well but take other forms than those that we associate with scientific endeavors: it may be the creation of an exhibition and then listening to the visitors' reactions; it may be the production of a book evoking a theme of enquiry obliquely so as to get fresh and unbiased reactions; it may be the production of a limited series of exclusive designed objects just for museums, to test reactions to the new ideas embedded there; it may be the creation of a dedicated, interactive web site.

And back to design-driven, in Verganti's book, as distinct from designer-driven: Italy may be famous for its design but has until recently no design education. A majority of Italian design is created by non-Italians, for companies without any designers on their permanent staff, and many of these designers have been trained as architects — architects who know that their work will stand far longer than the present market but also that it has to fit into a larger context. Good recipes for good design, and arguments for questioning the wisdom of market studies, are a must when opting for radical innovation in meanings.

Philips probes for the future of food

The Philips Design Probes program comprehends examining possible consequences of various long-range social trends and 'weak signals' emerging from the margins of society. In the case of food, this involved tracking and interpreting issues like the shift in emphasis from curative to preventative medicine, the growth in popularity of organic produce, implications of genetic modification, land use patterns in growing what we eat, the threat of serious shortages, and rising food prices. The result was three new projects, Diagnostic Kitchen, Food Creation, and Home Farming:

- Diagnostic Kitchen allows people to take a more accurate and, for each individual, relevant look at what he or she eats. By using a *nutrition monitor*, consisting of a scanning 'wand' and a swallowable sensor, you could determine exactly what and how much you should eat to match your digestive health and nutritional requirements at that particular moment in time.

- Food Creation has been inspired by the so-called 'molecular gastronomists.' These chefs deconstruct food and then reassemble it in completely new and different ways. The *food printer* would essentially accept various edible ingredients and then combine and 'print' them in desired shape and consistency, in much the same way as stereo-lithographic printers create 3-D representations of product concepts (cf. Chapters 8 and 10).

- Home Farming explores growing at least part of your daily calorific requirement inside your house. This biosphere home farm has been designed to occupy a minimum of floor space and instead to stack various mini-ecosystems on top of each other. It contains fish, crustaceans, algae, and edible plants, all interdependent and in balance with each other. Water filtration, recycling of nutrients, and optimum use of sunlight are all central to its appeal.

- In addition, Multi-sensorial Gastronomy explores how the eating experience can be enhanced or altered by stimulating the senses using the integration of electronics, light, and other stimuli. Developed in collaboration with Michelin chef Juan Marie Arzak, the three design concepts — Lunar Eclipse (bowl), Fama (long plate), and Tapada Luz (serving plate) — react to food placed on the plates or to liquid poured into the bowl.

The intention, as with all Philips Probe programs, is to elicit reactions and provoke discussion that can be used to further refine the ideas.

Adapted from Philips press release and web presentation

Endnotes

[1]Verganti, Roberto, 2009. *Design-Driven Innovation*. Harvard Business School Press, Boston, MA.

[2]Moultrie, James & Livesey, Finbarr, 2009. *International Design Scoreboard: Initial indicators of international design capabilities*. University of Cambridge Institute for Manufacturing, Cambridge.

[3]Lest my Scandinavian readers should think me ignorant, the Swedish Industrial Design Foundation, SVID, in collaboration with industry confederations, has performed massive interview rounds in 2004 and 2008, monitoring increasing corporate perceptions of design value.

[4]Lojacono, Gabriella & Zaccai, Gianfranco, 2004. The Evolution of the Design-Inspired Enterprise. *MIT Sloan Management Review*.

[5]Hargadon, Andrew, 2003. *How Breakthroughs Happen*. Harvard Business School Press, Boston, MA.

[6]Keynote speech at the International Conference on Design Research and Education for the Future, Gwangju Design Biennale, Republic of South Korea 2005: *Design Thinking. What It Is, Why It Is Different. Where It Has New Value.*

[7]The later adaptation: Owen, Charles, 2007. Design thinking: Notes on its nature and use. *Design Research Quarterly* 2(1), pp. 16–27.

[8]Buxton, Bill, 2007. *Sketching User Experiences. Getting the Design Right and the Right Design.* Morgan Kaufmann Publishers, San Francisco, CA.

[9]http://www.businessweek.com/innovate/content/dec2009/id20091214_823878. htm?link_position=link1.

[10]http://sloanreview.mit.edu/special-report/design-thinking/.

[11]http://www.openforum.com/idea-hub/topics/the-world/article/design-thinking-101-matthew-e-may.

[12]Berger, Warren, 2009. *Glimmer: How Design Can Transform Your Life, And Maybe Even the World.* Penguin, NY, NY.

[13]cf http://www.massivechangeinaction.virtualmuseum.ca/about/index.html.

[14]Lockwood, Thomas & Walton, Thomas (eds), 2008. *Building Design Strategy.* Allwood Press, NY, NY.

[15]Lockwood, Thomas (ed), 2008. *Design Thinking: Integrating Innovation, Customer Experience, and Brand Value.* Allwood Press, NY, NY.

[16]Esslinger, Hartmut, 2009. *A Fine Line: How Design Strategies are Shaping the Future of Business.* Wiley, San Francisco, CA.

[17]Lafley, AG & Charan, Ram, 2008. *The Game-Changer: How You Can Drive Revenue and Profit Growth with Innovation.* Crown Business, NY, NY.

[18]Brown, Tim, 2009. *Change by Design: How Design Thinking Can Transform Organizations and Inspire Innovation.* HarperCollins Publishers, NY, NY.

[19]Martin, Roger, 2009. *Design Thinking: The Next Competitive Advantage.* Harvard Business School Press, Boston, MA.

[20]Brown, Tim, 2008. Design thinking. *Harvard Business Review*, June, pp. 84–92.

[21]http://www.fastcompany.com/blog/fred-collopy/manage-designing/lessons-learned-why-failure-systems-thinking-should-inform-future.

[22]http://www.ub.edu/5ead/PDF/7/Durling.pdf.

[23]http://www.businessweek.com/innovate/content/jul2009/id20090713_332802. htm?chan=innovation_innovation+%2B+design_top+stories.

[24]Verganti, Roberto, 2009. *Design-Driven Innovation.* Harvard Business School Press, Boston, MA.

[25]Verganti has a nice web site for his book, collecting examples of new meanings brought about, here December, 2009. http://designdriveninnovation-collection. blogspot.com/2009_12_01_archive.html.

Chapter 4

THE ORGANIZATION, THE PROCESS, THE USER

The headline above is the opposite to the contents of this chapter. There is no single process, neither for innovation nor for design. There is no fire-sure prescription for how to organize for innovation or for design. And there is certainly no singular, only pluralities. The same, then, for the user: users it should be.

Yet, designers stress their ability to see from the *user's* viewpoint. Yet, design firms speak proudly of their *process*. Yet, corporations cannot do without *organization*. One reason is the analytical mindset dominant in business: to raise money, to weigh projects against each other, to 'optimize' the utilization of scarce resources, rational deliberations are seen as necessary. Yes, necessary they are, but they have their limitations: ponder Martin's pitching of reliability against validity, reviewed in the previous chapter.

The yearning for a proper innovation process description is so profound that historians of innovation research state that we have arrived at the sixth generation of innovation process descriptions. Only in the first few generations, the processes looked anything like linear, from... to... Later they became ever more 'chain-coupled' and networked,

63

with feedback and feed forward loops. Four-field 2×2 matrices and schemas share with linear models the feature that they are easy to draw on paper — often working only there, if at all.

Likewise with design. Brown, in his book, is at pains to explain that indeed they, designers, IDEO, have a process; to be a designer or a design thinker is nothing muddle-headed. So most design consultancies would proffer a design process, their own, preferably, often on their web site. These processes are most often linear, or described as such; reality, however, something else. In a wonderful compendium by Hugh Dubberly, "How do you design?", we find a systematic description of about 70 different process models[1] (sub-variations actually make the exact counting — inexact).

The reader would be eager to find a foolproof prescription for how to produce great products that were always guaranteed of success. We too! Needless to say, such recipes do not exist. The proof is in the pudding: much is dependent on factors such as timing, much a question of the competitive ecology when launching the new product.

We are dealing with design in so many senses — designing products, services, experiences, meanings, brands, value chains, and more — for companies large and small, diversified and niche-dedicated, global and local, manufacturing and publishing and servicing, that it would not make any sense to produce organizational schemas or design process descriptions. Instead, we will have to focus on our key elements: innovation and design.

Rather than thinking about success, perhaps we should think about failure. To fail often, to fail inexpensively, to fail so that we learn a lot. Thus, this is not a recommendation to willfully create products that are sure to bomb in the market, but rather the springboard for establishing criteria for ventures worthwhile to launch, successful or not. Taleb, in his important book *The Black Swan*, underlines that Black Swans — unusual, unforeseen events like successful breakthrough products — are by their very nature impossible to foresee or plan for.[2] Therefore, Taleb recommends that one abstains from huge bets on either undertakings perceived as "safe," which may only have the potential of turning into negative Black Swans, or on sectors that carry huge costs from collateral damage caused by negative Black Swans. Instead, the recommendation is that we make small bets on many potential positive Black Swans of which only the rare one will actually give any return, but then a huge one.

So those are our failures: small bets on Black Swans with an upside, with a learning experience thrown in for good measure. Do not over-analyze; it is more like firing before aiming. Design for a failure process. (Why is it not so popular to depict failure processes; no generations of them described?)

A designer strong on validity (following Martin's reasoning) can never prove in advance that an approach, an educated guess, will work. The strategy could rather be to map out *how* the guess can be tested rapidly and inexpensively, *how* also a failure would be an important contribution for an even better-educated guess, and *how* such efforts contribute to building a successful design activity.

Design thinking, if that is the word, or hybrid thinking should permeate the organization. It should be present in the stationary and the products, at the help desk and on the web site. In other words, as with P&G or Apple, it might require elements of centralization. Apropos Apple, Steve Jobs subtly states that they have no innovation process: well, "we hire good people."

Instruct on the differences between validity and reliability; train to accept the 'other side': design for that situation, re-design the situation. In creative problem-solving, à la the Creative Problem Solving Institute, an important step is asking: 'now that we have met those criteria, in how many and in what ways may management say 'no' anyhow, and how do we counter that in advance so there will be no 'no'?' Try to establish an area of shared values and ideas, and use that as the launching pad for understanding and seeking continuity from that ground, rather than a break. Share fully data and suppositions so that these can be challenged, built upon, and rearranged.

Claudia Kotchka, the revered head of Procter & Gamble Design, reported directly to CEO Lafley; she placed designers within the company's many business units so they could shape strategy directly instead of just designing how products looked. She educated P&G businesspeople about the potential strategic impact of design. She recruited a board of leading external design experts to offer guidance on how to make P&G a world-class design organization. She hired designers from outside the firm, while P&G is a company traditionally with internal careers only. She was allowed, no, she was supported doing all this (and more), so demonstrating the need for top management engagement. Likewise, we saw in the last opening case story how David Butler had to awaken Coca-Cola to the fact that it was a major design company.

Taking some leaves from Kotchka's book, and expanding on her principles, we might, e.g., consider having design as an item on the agenda for each and every meeting; include designers in every and all projects; perform an annual or, at any rate, regular design audit; arrange regular design seminars; and invite established and young and coming designers to work, demonstrate, speak, and discuss. And perhaps not just designers but other people from the creative industries. Producers of animated movies, say, artists into digital arts — or other arts... There might be internal seminars and workshops, prototypes and design experiments, web sites for testing design ideas and hosting a design discourse; there could preferably be shows or exhibitions of inspiring design, 'design tastings' where, like with wine tastings, reactions and discussions should be encouraged. The company could systematically partake in existing design competitions and arrange internal ones, recognize important contributions, arts could be hung on walls and rotated, everyone might be invited to sketch and contribute graffiti in assigned places. There could be a dedicated design space, a creativity space.

In addition to design auditing, we might suggest benchmarking design. A productive artifact is that of a *boundary object*, that is, an object that vividly represents the object and objective of a design project, creating a direct, pervasive, convincing, and inspiring presence of it, furthering discussions and reflections. The Web can be relied upon too, for example, for multi-location settings.

In a study of group versus individual decision-making, military experts were given intelligence data and asked to analyze troop movements. The experts scored badly against actual troop movements, with only an average of seven percent correct judgments. Then, every expert received the analyses of his peers and gave a new assessment: now, they reached an average accuracy of almost 80 percent. No new information, just a lot of new ways of seeing. There is great wisdom in crowds.

"Creative friction" generates advantages when a design client engages multiple design firms — a 'second opinion,' so why not a third? Several design firms mean access to a larger network of capabilities, solutions, and experience with earlier projects resulting in a larger design scope, with varying benefits depending upon the particular project — and for design consciousness in general. Another type of friction, or challenge, is when one is engaging to design something unusual, and not within the business, but to obtain experience, understanding, and competency. An example is to stretch projects, like a guitar without strings (laser light beams instead); Porsche doing sunglasses, luggage, and kitchen knives (perhaps not 'stretch' any longer); or a car design studio relying upon its car chair expertise to do regular furniture for sitting.

The design discourse could prove pivotal. Network members should be encouraged, and rewarded, for contributing beyond contracts. The discourse network should be maintained and developed, as Verganti[3] suggests, preferably inventoried and perhaps gauged with some kind of metrics, qualitative if need be. Sometimes the network is multi-tiered, that is, it is a design chain. Partners should be involved in experiments, and experiments performed as often and quickly as possible. It is important to recognize that those networks encompass much more than just designers — architects, publishers, journalists, people from the media and creative industries in general, scientists and more; knowledge brokers a category of its own — and that it is truly global. The discourse is seeded with various initiatives, cf. also the Alessi activities which are described later in this chapter.

Sketching and probing: cultural prototypes, a book, a web site, a community, an exhibition — testing new meanings by bringing them out by various media to foster the discourse and the feedback; an interim prototype or mockup.

Christopher Frayling has established an important distinction between three different kinds of research related to design. Research *into* art and design is about, say, art history and the evolution of styles, and understanding perceptual and cognitive aspects of design and art. Research *through* design is when design is employed to gain a deeper understanding, e.g., into materials or technological applications so that research is really in the service of design; this equates with efforts to establish Martin's validity. The third is research *for* (or *in*) art and design, where research and design are intimately coupled and intertwined: design becomes a component integrated with research process and results.

Philips is a company that has taken design to its heart[4] (if a company can be said to have one!). Here, they talk of research for design as important to be on top of designing, that is, for example, successfully detecting and following significant cultural trends. By contrast, research into design equates with new methods and design languages, another important category. The third category, within Philips tagged with the preposition *through*, is when design is used as a research tool in and of itself.

Such research opens up new vistas that may rather be seen as an entire network of options, which may not instantly be translated into business propositions. In the 1960s, the Swedish businessman and business philosopher Gunnar Ehrlemark developed what he named SISTEM, with three horizons corresponding to three different operations: orientation, planning, and programming. Somewhat similarly, Philips design thinkers apply three different horizons: creating viable options (broadly a parallel to Ehrlemark's mapping); developing new business (going somewhat further than planning); and defending and extending existing core businesses (adding to programming).

In 2000, Philips decided to design a research program to explore natural, intuitive ways to affect the environment with multi-sensorial stimuli to create an atmosphere of rest, intimacy, imagination, and play. The aim was to learn how to design experiences. In one experiment, sleep was the object, and 'sensing' linens (for monitoring effects), camera, projected images, and an alarm clock were among the instruments shaping the environment. Results could be translated into the viable options horizon: mundane life could be differently and much more richly experienced. This mapping was used to generate possible families of applications and basic patents. Teachings, ideas, and techniques developed in the experiment were introduced into healthcare facilities, such as demonstrated in the MRI application in our case story prologue. In order to go to the next horizon and approach concrete implementation, in a vast corporation like Philips, collaboration had to be secured, cast in projects run by and for partnering consortia.

Results from the sleep experiment materialized in what was called Ambient Experience, an entire environment now introduced in health care. First out was the magnetic resonance business unit in 2003. Jeroen Raijmakers, Global Design Director of Healthcare at Philips Design, stated that,

> "There is only so much you can do during product design. If you want to make the experience even more complete, then you also have to design the context in which it is experience [...] We called it ambient experience design because it embedded light, multimedia and interaction with users in an architectural context. We took a close look at the phases patients have to go through during a radiological exam, from expectation to memory. Needs of patients and professionals during waiting, preparation and the actual scanning procedure were identified and translated into solutions. We came to the radical conclusion: the whole radiology department should be redesigned."

Philips had had to change its way of designing, what it was actually designing, and how to sell and implement ideas created — not so easy!

So this is the second phase or horizon: a more concrete development effort, with strong relational components — clients had to be assisted in working differently, but existing demos helped clear the way. The next challenge became the clients' very own in that they had to change their ways of operating.

When the Ellemtel development company was given the task of designing an electronic telephone switch, what eventually became the tremendously successful AXE system, they were in a hurry: Ericsson seemed to miss too many orders. Still, the development team took the time to ponder the fundamentals and to research both the state of the art, potential development trajectories envisaged, and also a bunch of ideas classified as outrageous or marginal that might have a bearing upon the project if it was ever realized despite bad odds. The team took the time to get the system structure right and the right concept first, with some clear principles to always guide them. No system yet, core principles only: options horizon, mapping.

Only after that fact, it dawned upon the Ellemtel people that they had stumbled upon yet another principle. Different skills were needed at different stages of the project with different ways of operating — the head of the project handed over his responsibility to a person to complete the development work, and he, in his turn, handed over to others to do the actual concrete translation to business and industrial practice (because of the differences between the actual systems developed, parallels with the Philips experience can be drawn only so far).

Likewise, at Philips, the three horizons have been found to require different competencies. The new options must be given the chance to find their true places in the market, much as the first head of the AXE project was surprised that his system was ideal for cellular telephony, a market non-existent when he and his team designed the system. He had, unwittingly, in a generic way claimed a large application space, larger than he knew because he was thinking in broad or generic ways ("the most flexible system possible," see also Chapter 9).

An important aspect of the design discourse is to keep tabs on emerging trends and fashions in society. There are many ways to attempt to do so: finding hidden trends in newspaper clippings and advertisements, observing trendsetters (teenagers?) in their chosen settings (central Paris, a park in Tokyo), bellwether states (in the US), as well as subscribing to trend letters (the box offers a sample of such, available through the Internet). As an example, Philips Design has established a trend group, the VTA group, which together with an

Here are a number of web sites attempting to follow trend developments:

http://www.trendwatching.com/

http://www.trendhunter.com/

http://www.influxinsights.com/

http://www.psfk.com/

http://www.fashion-ation.net/

http://www.springwise.com/

http://www.thecoolhunter.net/

www.shapingtomorrow.com

www.fringehog.com

www.technorati.com

www.futureconceptlab.com

www.geniuslocilab.com

www.news.agendainc.com

mooch.info

www.proteinos.com

international team of trend watchers for a particular range of products, named Essence, identified a target market of design-interested people with 'traditional' tastes, who are a bit conservative, classy, traditionalists (Philips calls them Traditionals). The VTA group wrote a booklet to describe the traditionalists. Products should feel calm and serene, and not change too often. Two sets of color combinations were chosen to attract this niche, one off-color white associating with the quality of ceramics, expressing careful craftsmanship, the other an appealing grey.

The greatest innovative value is realized when designers are engaged early on and permitted to participate all the way through the process, including being engaged also with last-minute changes or practical adjustments to the product. Advantage from designers' participation accrues best when designers — be they external or internal — are given sufficient clout.

Alberto Alessi, Ernesto Gismondo of Artemide, and others dedicated to radical innovation through design do not set their trust in market studies. Apart from feeling the winds of what is contemporary, or, rather, can be created to feel contemporary, they perform research, that is, various forms of tests and probes to generate reactions. One facet of the design discourse, apart from engaging designers and trend watchers and journalists, is to involve consumers in a process of co-design. Such efforts need their own setting, a space designed for the purpose, sensuously inspiring, with preparations and instructions to match the task.

The approach used in practice — and hence the patterns and concepts chosen for further development — may, of course, be dictated by the client or other stakeholders rather than being the designer's decision. Half of the ideas for Alessi's products come out of the firm's initiative, prodding designers; the other half when one or the other of the two hundred designers that work for the company contact Alessi with an idea to discuss — first by phone, maybe, then by e-mail (other designers, 'non-regulars', also make many suggestions but rather rarely ones that become accepted). The discourse *initiatives* encompass 4–6 workshops per year, often with young designers from universities and industrial schools across the globe as well as recommendations and tips from journalists well acquainted with Alessi. Furthermore, there are *design explorations* of the type Verganti heralds (see the previous chapter): for an emblematic domestic object, a group of designers, most often architects, are given total freedom to design without any constraints; *competitions*, used by other companies, are less frequently used by Alessi. From more than twenty products designed and prototyped, a handful are explored for mass production — a long-term and costly undertaking.

Alberto Alessi compares their mode of working with that of Picasso in the sense that the artist followed his intuition, not any thought of a target group or market niche. Alessi sees no reason for risk reduction, rather for finding the right balance between risk and opportunity potential. He uses the term 'mediator' for what his firm does: mediating between expressions of product design on one hand, and customers' dreams on the other. The task is that of orchestration, like that of a movie director.

Relational assets: it is established practice to perform competency reviews, to nail down where new opportunities may arise from un-thought, un-established new combinations of existing competencies. What about a relationships review or audit, mapping the company's various networks to see how these may be intertwined productively? And needs for complementing?

As stated initially, it would be entirely inappropriate to suggest any standard process description for a non-existent design process, and equally inappropriate to proscribe a certain set organizational solutions. Rather, we have discussed a number of recommendations for enhancing an organization's design awareness and agility:

- Develop, maintain, refine, and permeate a common language, securing a profound understanding of dimensions of design for innovation
- Try to develop metrics for important aspects of design undertakings, relevant to your company and context, realizing that such metrics may largely be qualitative and that much of the value lies in discussing how to design it

- Look into and try to improve your failure process; granted, the word 'failure' is tainted and one aspect of your language might be to substitute it verbally with something more associated with trying, experimenting, or learning

- Understand and communicate the differences between reliability and validity, and try to establish regular ways to balance them

- Inventory individual qualities to learn more about the 'T person' qualities, cultural and other diversity, and personal networks

- Take note of the various recommendations on how to make an organization more design-prone and conscious, and implement them as they fit

- Make sure to benefit from the wisdom of the crowds, always having people share their judgments though these may have been prepared independently at first to avoid groupthink

- Make sure to design an adequate, dynamic design discourse, one to grow as the scope for design and the scope and network for the discourse grow; inventory design relationships and explore different ways to research relying upon the discourse, seeding it with adequate probes

- Take always into account the different horizons, effort, dimensions, and steps involved in design-inspired innovation. Do not forgo the vast potential with the early options horizon that, however, is just necessary but not sufficient, and which needs several more steps to lead to business proposals, while intellectual property should be considered already at the outset.

Who are the end users: when, where, and why?

A central tenet for industrial designers is to take the end user's view; designers "who listen with their eyes" as Claudia Kotchka proposes.[5] A host of tools and methods can be listed for understanding the prototypical end user, who, for that matter, may be someone else than the customer, the one paying for the designed object, product or service. Employ focus groups, anthropological and ethnographic 'deep dives,' interviews, diaries, frustration maps, annotated or registered with video cameras or electronically, GPS and accelerometer equipped monitors, touch-points registered... extraordinary users or outliers or odd behavior for breadth and discovery, few users known intimately for thorough knowledge... lead users, communities of enthusiasts, tribes like the Harley Davidson riders' community...

By definition, innovation involves novelty. Market research and interviewing are of limited value for knowing whether there is a market for something still non-existent. Is the end user necessarily well known or defined, or perhaps just a myth? Asking 'the market' about potential features risks resulting in 'featuritis' and unnecessary 'complexification.' Nothing beats actual testing of the market with real products. Tokyo's Akihabara district is a showroom for novel electronic gadgets, sometimes given credit for Japan's success in such goods. Good salesmen can gauge hidden needs and hints about what would make a product attractive when given the incentive to communicate such signals. With the Internet, it is possible to handle and supply panels and test-markets with products still pondered. Perhaps one should consider Beta version products, like with software and web services? Before arriving at products to test, however, the seeds of innovation may have been planted — through unusual insight obtained far from the road most traveled.

When prototyping a product, several prototypes serve to arrive at a final, appropriate and attractive solution. Prototyping end users involves casting an ensemble of them, comprehending the vast majority of typical users of the eventual design, and also their varying user situations — and, next, the non-typical ones. Mini-biographies of end users, so-called *personas*, may serve as potent tools for getting under the skin of them, to try to understand how each of them would react with the designed object.

Dividing the world in suppliers versus users is one way, and also conventional. But may we not change our attitude and approach by viewing them as users rather as an audience? Or as participants or members of a club? Or as visitors or, as in some theme parks, as guests? What change in approach does it bring to see them as crew members?

The user interface is an important factor and may involve trade-offs when making it adaptive to individual or situational needs. Situational: what about the need for learning to use the designed object, what about its affordances, its functionality available without instructions, just following one's intuition? What is the emotional journey that the user experiences when meeting and interacting with the design (and before and after; Manguel points to the fact that re-reading a book, knowing the plot, is then an entirely new experience)?[6] What is the experience of the first encounter, and what at the 100th? What are the differences between attracting the customer to buy a product or a service in the first place, and then continue to use it, to like it, and to recommend it to others? Enjoyment and memories are the two factors that cause people to hang on to certain products, and after an initial love period, attachment falters, to be revived later for favorite objects: lessons for after-sales support or communication. Are there differences between habitual users and those using the object only seldomly?

It is often helpful to list and organize such factors in matrices, flow schemata, and mind-maps. The different personas seen as involved may have partially different needs and weigh the importance of the design of different touch-points differently. And what information, what ancillary products or people are associated with each touch-point?

One person plays several different roles during a day, switching between them: voluntary scout leader, parent, attorney, home-owner. Would the goals pertaining to the designed object differ, or the weights associated? What if the user has not slept well, or happens to be calm, excited, or nervous? Contexts will vary: the same web page at work, leisure, and in the smart-phone on the move is not the same, really; but all too often, products are designed without any systemic view, leaving the product design isolated from its larger context, and thus function-ality. Looking at the importance along touch-points generates a list of priorities. Generating options for dealing with problems or frustrations should result in costs tagged to resolving or easing problems: a way to arrive at more attractive solutions with lesser downsides.

The context and the user's need, in relation to your designed object, can also be seen to be associated with user behavior. If alternative behavior is difficult or shut out, then there is no or almost no behavioral choice, and behavior is severely *constrained*; a particular type of constraint has been called *tunneling*, such as when you are guided through the installation of new software. An *enabled* behavior opens an easy way for the user to take the prepared route, without, however, shutting out alternatives entirely. The design can be *adaptive* so that the user directs it to allow for a behavior the user prefers. And then, there is the possibility of *motivating* the user, through various inducements and incentives, education and resort to established habits, routines, or templates.

Seeing, listening, and posing basic, 'stupid' (frog design's expression) questions are key in several of these methods. Many years ago, a Japanese company could not understand why it was just impossible for them to sell their well-designed rice cooker. A focus group discussed rice cooking, and only by re-listening to the taped discussion very closely, a diligent listener heard the revealing phrase expressed in passing: how could one be sure that the cooker was clean after the previous meal? Listening should preferably include attention to the qualities of speech, such as rhythm, timing, hesitance, etc., and also to non-verbal cues such as gestures by hands, head, or body;[7] such cues (brow raising, drawing back in a chair, compressing lips, gazing away, etc.) may differ in significance between cultures.[8] There are many ways to elicit reactions, such as asking people to choose mood figures rather than words; sort cards; or rank reactions to objects in different categories.

A G Lafley, always keen on learning first-hand how Procter & Gamble's products were used in real life, went down to laundry basements in Tennessee and Kentucky to observe, nay, to experience intimately how Tide washing powder was applied. When consumers were asked, they always gave Tide cardboard packaging rave reviews. Amazed, Lafley saw no single woman open that package with her hands but with screwdrivers or nail files or scissors or whatever. Why? They did not wish to risk breaking their fingernails. "We thought our package was excellent because they were telling us it was excellent" (hinting that much else was likewise told as excellent).

Lafley made that real life visit (and many more). Customers also pay visits; a study revealed that half of home electronics products being returned to the store did not feature any malfunction but for the most important one: consumers did not understand how to operate them (actual malfunctions are, of course, another source of information). A new 3M product started life with a complaint call to customer care. The 3M attendant did research online, suggested a solution, filmed it in a video which was put on YouTube and re-contacted the customer to learn whether this was what he was asking for. Oh yes — and the product got to life. A design discourse?

To learn about behavioral triggers and inhibitors, it is, of course, possible to set up laboratories, sceneries, or environments where subjects are enticed to react on stimuli that are considered important to a design project. Certain objects or certain design elements may then stand out as defaults that should be left as such; components or motions or signs that have become habitual and part of users' usage shorthand. On the other hand — there is always another hand, no? — design, too often, is about the front, the surface, what we see, but not how end users react to the unwieldy or chaotic interior of a product, or the back, rear, roof, or the underneath of a nice product or architecture?

We have mentioned the idea that synthetic archetypal stand-in personalities, *personas*, may play a constructive role in understanding end users. There are many good sources that instruct on how personas may be designed — here it would be appropriate just to provide a brief overview.

As with other research methods, it is important, when obtaining information to base personas on, to be wary about whether what people say actually reflect what they feel or mean — even though they believe so when saying it (P&G's Tide carton!). Give personas names and photos, make them live, and describe their history and their environment. Age group, income, wealth, and other demographics go into a persona description. Culture, job, and educational background should obviously be included, and then attitudes, time management behavior, constraints, motivations, social mores, frustrations, and critical and inquisitive questions should likely be phrased in relation to the product or service we design — and expectations on it. Personas are made with a particular design task at hand, including what a purchase process might look like; they are not freewheeling. The efficacy of personas has been questioned, and a cure-all they are not; what is? Research comparing design results with and without using them tell of a clear difference in favor of using them.

For cultural differences, there is the Inglehart-Welzel map in two dimensions, traditional to secular and survival to self-expression, respectively.[9] Most designers would be found towards self-expression while survival may matter to a large part of world population and 'end users.' In addition, there are cultural differences in habits that matter, for example, making for different demands on assistance systems for car drivers in Sweden versus China,[10] or ways of perceiving web page content.[11]

Users' objectives may be categorized as functional, emotional, social, and seeking for meaning. These are not necessarily independent. Emotions may be activated by beauty and aesthetics, but also by good functionality and rewarding social effects: fitness to social standards, for example. In addition, surprise and interest are emotions that may be evoked; generosity, an action. At the functional end, basic physiological facts must, of course, not be ignored; ergonomics to the rescue, while variations between users should be taken into account. At the other end, meaning and emotions may meet. But, of course, price, or value,[12] comes into the picture — much dependent upon the context. Communication and community are involved in shaping the product's/service's space or ecology: other people's perceptions do matter.

The best buy-in of all is when people integrate into their lives that which was designed, shaping other objects or actions around it, co-creating the fuller context. What are the stories they tell about your products? What exaggerations do they forward? More generally, what adjacencies, hardware and software, and maybe social action or connections, are involved? What constraints?

Ergonomics can and often should be tested, as should user experiences and user interfaces. Customer feedback is, of course, a bit late but can be fed forward for the next product generation: complaints constitute a resource. Pre-paid postcards with easily answered questions may help generate feedback. But customer complaints can be seen as a resource, without any postcard-prodding involved.

When designers Barré & Associés got the task of designing children's toothbrushes for Elida Fabergé, they started by listing the negative aspects of the then current market's children's toothbrushes. These were found to be non-existent in the sense that they looked just like adults' brushes, while 3–6 year-old kids are uninterested in brushing their teeth — playing is attractive, though. So a new playful design was conceived: a standing up brush, looking alive, a playmate. Success: the Signal brand market share jumped from 8 to 42 percent. To keep that market share, more resources were allocated to design, and now more than a handful of kids were actively involved in a development effort which consisted both in the children reacting to various design proposals and them designing 'lively toothbrushes' on their own. Resulting prototypes were tested on mothers and 3–11 year-old kids, resulting in the next generation of Signal Croissance toothbrushes. The next step was to broaden the scope to develop a set of accessories for toothbrushes, including a box for baby teeth, serving also as a stand and protection for the brush. Fabergé was now content, but the designers were not. They found the brushes to tip over, and developed a suction disc, a solution that they got patented, and which Fabergé adopted, paying the design firm for each toothbrush sold — now available in two sizes for different age groups.

Ethnographic studies are performed not in a lab, but in the field. Functional or value analysis gets to tradeoff between functions, value, and cost/price. The pure physicality of the user environment is important to consider, to observe, to take into account for ergonomics, for user interfaces, for sheer practicalities, for timing and procedural aspects, and for the possible development of add-on objects and services.

Girls like people, boys focus on things. Women underestimate their intelligence, men overestimate theirs, while women are better at language and emotions, men at spatial skills.[13] To park a car in a narrow slot between two other cars took men an average of one minute; women three minutes — cars should be equipped to cater to both sexes. Volvo got much favorable press for having women design 'the ladies' car' with a number of ingenuous details never seen in a car before. A map maker had observed that women read maps in the direction they were moving so printed the map almost the same on both sides; almost, since the names were upside-down. They sold 15000 copies to women, a handful to men.

Men and women use different words indicating different preferences when judging designs. Both women and men prefer objects with characteristics such as compactness, slimness, and cleanliness, but men associate with keywords such as solid, ergonomic, and relevant — words that women do not use at all. Perhaps men prefer products that seem sophisticated and reliable. Women apply words like curvy, flexible, decorative, and neat — a preference for organic forms.[14]

What meanings would the user feel are conveyed by a product or service? A complex one might be termed an accomplishment, and what has been accomplished might be seen as a creative contribution, an experience of enlightenment, or a fulfillment of duty, a service to or from belonging to the community, possibly also as redemption. We deal in this book at length with beauty, often partnered with truth and justice as fundamental yearnings; wonder and awe combine in producing the sublime. Among Maslow's fundamental needs, security is one. Perhaps freedom as a concept has not always been as highlighted in history as now, but then it arrives, or we are deprived of it, in several flavors. A sense of harmony and oneness may be seen as associated with community, but not necessarily.

To change the perspective: what might best empower the user?

To learn from Verganti's design-driven innovation, the proposals that Alberto Alessi offers his audience have been described by himself in several interviews and in Vergantis' book[15] as being characterized in four dimensions, which all can be rated on a scale from 1 to 5.[16] For markets where the Alessi company has profound experience, the formula is quantitative with high precision. Two dimensions are traditional ones, to be expected, and Alessi does not regard them as the ones setting his company aside: function and price. Here, it is not too difficult to establish what works, and what tradeoffs to work with. One of the two more original dimensions is SMI, which is shorthand for sensory, memory, imagination. This is gauging whether people interject "how delightful this object is." The other shorthand is communication language, CL, which stands for the extent to which the product conveys values, life styles, status — in one word: personality. This is sometimes described as the meta-product.[17] As low a composite score as 12 is deemed very risky whereas higher figures indicate increasing potentials for capturing larger audiences. Remember: Alessi is risk-prone. SMI is the factor Alessi picks as the one hardest to nail.

Do learning styles matter?
And do they exist?

People, that is users, may also be characterized by having different learning styles — may, because the whole idea of learning styles (there are more than 70 suggested) has been criticized. With that warning, we make a sampling of some of the more frequently discussed. Perhaps most straightforward is the distinction between visual, auditory, reading-writing, and kinesthetic/tactile learners. Another one is for cognitive ability: analytical, creative, practical. Kolb offers a two-by-two matrix with concrete experience versus abstract conceptualization on one axis, active experimentation versus reflective observation on the other. That gives him *'accomodators'* for concrete and active experimenting (here designers are overrepresented, by the way, compared with MBAs and engineers); *'divergers'* for concrete experience and reflective observation (designers underrepresented); *'assimilators'* for the abstract and reflective observation; and *'convergers'* for abstract concepts and active experiments: to ponder ideas, then test them (designers less represented). Kolb's model has also been restructured into a stage model. Gregorc had the matrix organized after abstract–concrete and random–sequential, getting four combinations and learning and searching strategies. Yet another model would go after personality styles, where *'monarchic'* learners prefer focusing on one thing at a time, *'hierarchic'* learners prefer a hierarchy of goals, while *'oligarchic'* learners allow for competing approaches, and *'anarchic'* are flexible and idiosyncratic. Another dimension entirely is the fact that with changes in society, different generations are attuned to learn differently.

Exovation — the flip side of innovation? And — exodesign?

What are the obstacles to innovation? Numerous, to be sure. One or several consist of concerns, investments, and activities that occupy space — organizational, cognitive, economic space draining innovation of necessary resources.

A company credited with profound concern for innovation and thus featuring important organizational features for furthering innovation is 3M, the corporation of Scotch Tape and Post-It Notes® fame. Academics and journalists alike have fallen for their charms, though criticism has also been voiced.

One of their innovation mechanisms is found in the establishment of precise criteria for renewal. Each and every business unit has to get a certain *share* of their revenue, thus a quotient, from products that are younger than a certain amount of time. So X percent of revenue must be generated from products that are no older than Y years. (The reason for not mentioning X and Y is that they have been adjusted at least once to reflect a need for higher speed, and I am not certain what the current levels are.) The criticism I mentioned takes issue with a quantification that may be subject to manipulation, a risk when applying rules blindly and rigidly. So wisdom and common sense must prevail.

The obvious route to achieve the required share is to get new products going, to innovate. But do recall that we talked about a quotient: new products enter in the dividend, but to get the ratio, there is the divisor too. In this case, we have seen it consisting of revenue generated by the 'old' products, aged more than Y years. So a complementary way of achieving the demanded quotient would be to reduce the divisor.

It is here that we find *exovation*, the power of GETTING RID OF (some) prevailing products… Behind the 3M dictum, we discover the profound insight that there are older mainstay products that are just about marginally profitable, if even that, and not just forgotten — living dead — all the while requiring attention, engagement, involvement, management resources, routine but not for-free upkeep such as storing, catalog updates, inventorying, information in data bases, and bookkeeping. All activities taking focus away from the creation of novelty, all ties to the past, all routines that create lock-ins — requiring time, attention, interest, engagement, thinking.

Another large American corporation showcases another type of exovation: the organization in its future. They have recognized that the organizational structure of today reflects and thus tends to permeate the past; while the only thing that is entirely certain is that the future will be different. Therefore they have created a skeleton alternative organization for the future as anticipated, an organization with its own resources for creating novelties, for innovation. Since the guesses about the future are certain to be wrong, or not entirely right, there are free resources for projects that do not fit neither the current nor the skeleton future organization, but that still merit being treated as lottery tickets.

When Jimmy Carter was elected president in 1976, his program was much about better management. One particular management practice, fashionable then, which he had practiced as governor in Georgia, was zero-base budgeting, and this was highlighted in his campaign. The underlying philosophy is, in our new term, about organizational exovation: in each and every organization, there is an inertia that makes it continuing all procedures just as in the previous year — with 4 percent more in the budget or maybe a 3 percent reduction on the margin. The zero-base budgeting recipe demands starting all over from the very beginning, relying on first principles to really understand what is worth paying for, what merits resources, what the trade-offs really are. As with so many radical recipes, this would be just too demanding to be followed to the letter, overall and every year, but there is at the bottom an important exovation proposition.

What might a program to develop methods for furthering exovation look like? An obvious route would be to design several more practical means to translate 'more for the future, less to the past' in practice. A natural approach for a researcher is to suggest that we collect, organize, and promote what we already know about how existing projects get in the way of new ones — like Thomas J Allen has demonstrated how dangerous it is to decide on one single development trajectory too early into a project.[18] Above, we saw a list of a number of ways in which existing products contribute to inertia; that list might be enlarged, systematized, tailored to specific activities, and its various components assigned individual weights.

Should we also think about *exodesign*, getting rid of stale design? Even design classics may eventually become tired and constitute more of a hindrance to new success. Like with exovation, it makes for a new way of looking at design.

Endnotes

[1]Free download at http://incsub.org/soulsoup/?p=1081 or http://www.dubberly.com/articles/how-do-you-design.html.

[2]Taleb, Nassim Nicholas, 2007. *The Black Swan*. Random House, NY, NY.

[3]Verganti, Roberto, 2009. *Design-Driven Innovation*. Harvard Business School Press, Boston, MA.

[4]The following several paragraphs rely heavily on Philips web pages and particularly on S Kyffin & Gardien, P, 2009. Navigating the innovation matrix: An approach to design-led innovation. *International Journal of Design*, 3(1), 57–69.

[5]Strikingly, 2009. Literature Nobelist Herta Müller uses the very same phrase to describe how people acted under the oppression in Ceauşescu's Romania, with its listening walls, homes regularly ransacked, and people purposely hurt or killed (Müller, Herta: Der König verneigt sich und tötet). Carl Hanser Verlag, München & Wien (2003); originally from a lecture she held within the Tübinger Poetikdozentur 2001 — read from the Swedish translation 2009.

[6]Manguel, Alberto, 2008. *The City of Words*. Continuum, London.

[7]For different dimensions, see http://www.ijdesign.org/ojs/index.php/IJDesign/article/view/313/164

[8]For human movements, there is an annotation system, developed for dancing, called Laban after its inventor.

[9]From World Values Survey, http://www.worldvaluessurvey.org/; the map itself at http://margaux.grandvinum.se/SebTest/wvs/SebTest/wvs/articles/folder_published/article_base_54.

[10]http://www.ijdesign.org/ojs/index.php/IJDesign/article/view/354/165.

[11]http://www.ijdesign.org/ojs/index.php/IJDesign/article/view/267/163.

[12]Boztepe, Susan, 2007. User value: Competing theories and models. *International Journal of Design,* 1(2), 55–63.

[13]Adrian Furnham in http://www.newsweek.com/id/101079 and http://www.ucl.ac.uk/lhl/lhlpub_spring08/04_150108.

[14]http://www.ijdesign.org/ojs/index.php/IJDesign/article/view/71/76.

[15]Op cit.

[16]See also, e.g., http://www.fastcompany.com/design/2009/featured-story-alberto-alessi.

[17]Linn, Carl Erik, 1990. Metaprodukten och det skapande företaget. Liber, Stockholm (in Swedish); an English language briefer, more recent document can be downloaded here: http://www.metamanagement.se/.

[18]Allen, Thomas J, 1977. *Managing the Flow of Technology.* The MIT Press, Cambridge, MA.

Chapter 5

BEAUTY-DRIVEN INNOVATION?

The very first sentence in *Design-Inspired Innovation* proposes a design-inspired product to *delight* the customer. More generally, what might delight the customer or the user? Why do people buy a product, or a service? Price. Quality. Performance. Price/performance. Technology and attractive features. Customized. Adjacent services.

Yes, possibly all of the above. You might think that commodities would qualify for the 'price-only' category, but not necessarily so: Chiquita has succeeded in establishing branded bananas. The arguments may shift depending upon product or service, and with customer or user; sometimes the individual buying or influencing the decision to buy may be someone else than the end user.

Price may come down as an effect of more efficient production, better logistics, lean thinking, economies of scale. Quality may be designed into the product as well as into the production process and indeed the entire organization (or network). Tools and methods such as statistical quality control, kaizen groups, and fine-tuned mechanisms for listening to the customer and user may all play a role. Performance may depend upon technology features but also services associated with a product, or the user interface offered. New technology may offer vistas to the renewal of existing products, enhancing their performance, but also sometimes allowing for entirely new products, product categories, or, though rarely, new industries.

Quite naturally, every new or, rather, newly focused factor of competitiveness follows a learning curve, with rapid advances and return on efforts at first, then a concomitant slowing down. The pioneers have the upper hand initially, reaping the benefits of the new discovery, suffering some pains of the untested also. Eventually, every competitor applies about the same toolbox — applying 6 sigma to achieve high quality and relying upon the same basic technologies to the extent that performances and features converge.

Management methods follow similar development curves, demonstrated by the trajectories of, e.g., future studies, benchmarking, business process re-engineering. Some methods and some ways of competing are easy enough, if not to copy, at least to adapt and implement successfully; others, perhaps, impossible because they depend upon deep structures such as corporate culture and profound systemic features ('an organization's DNA'). Those systemic features may be seen as being designed in a very basic sense; corporate culture cannot, however, be designed but has to be regarded as intrinsic.

New dimensions and new tools for competing are eagerly sought after. One such feature not listed with price, performance, or quality above would be beauty.

Beauty has been seen as a fleeting quality, avoiding hard definitions and measuring devices. Beauty, we are often told, is a subjective quality, in the eye of the beholder. Beauty is oftentimes subject to fashion, thus fleeting also in time. But not always: there is, moreover, everlasting beauty, neither subject to fashion nor to trends — some would call it classic.

This is not the medium (nor the author) for a philosophical treatise on beauty. But it will be the place for summing up some pertinent arguments and views on beauty, with the aim of provoking a discussion on how these may be applied to the readers' businesses and, indeed, design work.

In your business, what would qualify as beauty? Who would be qualified to judge? What examples are there? Which are the classics, and what would be examples of fashions and fads?

Try to decompose your concept of beauty into a number of dimensions! Are geometrical elements of importance? Color? Visual balance? Balance, generally? Proportions? Harmony? Symmetry? Sound? Smoothness, shininess/reflectivity, texture, pattern, curviness, simplicity, usability, velocity, naturalness, or modernism? What about tactile sensations, like when opening and closing a door, to a car, a stereo cabinet, experiencing a comforting precision? The feat of obtaining the same color impression from a metal surface and the adjacent wood surface? The surprise in seemingly impossible body curves of a new car? Elegance, in the movements of a robot behaving like a ballet dancer, as yet an example?

Steve Jobs' phrase is well known. He wanted the computer or the iPod to be "lickable." What would be the most appropriate attribute for you and your business: likable, slick, mind-blowing, attractive, vow!, awesome, striking, or what? Try them out! Our lists of attributes by the end of this chapter certainly provide you with a large catalog of attributes, but anyhow these lists are in no way exhaustive; try to amend to them, and, again, try to see how the various attributes might apply to your own activities.

Innovation is much about bringing together as diverse components as possible, thus, bringing design views and designers to the party, widening the scope and the potential for breakthrough creativity. The words mentioned above, and the questions linked to them would not always be the ones associated with ordinary product development.

The words in the lists at the end of the chapter may, at first, in some cases seem synonymous, but at closer scrutiny, they constitute synonyms only partly. Given that word meaning is in the mind of the beholder and tinged by associations created through personal history, culture, and language, would not 'beauty' be an ethereal quality, while 'striking' is more immediate and abrupt, 'mind-blowing' even more so, with a larger dose of surprise to it? Try out what distinctions in meanings you might deduce while pondering those words, and the lists you have extended.

Most often, we do not design for showpieces, but objects to be used. Therefore, beauty must be seen in action, in interaction between user and designed product/service. Learning to use, or using a product, may be seen as, if not beautiful, at least elegant and attractive: enjoyable. What is necessary for the object to be utilized should also, preferably, be irresistible, and positively so.

Most people would have some favorite objects that they would prefer never to part with. MIT sociologist Sherry Turkle calls such specimens of technology, revered by an individual, 'evocative objects' serving as springboards for identity, security, reflection, or thinking. A doctor may love her stethoscope, serving as a badge as well as a tool; a writer might cherish a special pen. Turkle says, "we think with the objects we love, and we love the objects we think with." She suspects that most of us have some kind of technology that acts as our touchstone: what's yours, what's your customers?

If beauty is in the eye of the beholder (though is it, really? More on that question later), who is the beholder? When General Motors overtook Ford as a car manufacturer in the late 1920s, it was on the realization that the market was no longer homogenous (if it ever was) but rather that different customer groups were attracted by different types of cars, in different price brackets.

So perhaps you would have to consider whether to appeal to some kind of average sense of beauty, or to offer different models to different groups, not based upon technical features but on different tastes for beauty.

As the old saying goes: what gets measured gets done. How to measure beauty — in your line of business? What constitutes beauty?

Now take a look at the list of negative attributes: what is ugly, what attributes apply, which are more moot? What do you learn from testing and discussing their relevance, or irrelevance, to your products and services? The same goes for the list of attributes that do not easily fall in the categories of ugly or beautiful: try to complement it, attempt to use it as a litmus test to discern what qualities are applicable for your activities.

Now take another tack. What if your company was a sport, an animal, a flower? Your product? Why would it be that particular sport or animal: what does it say about the qualities of it? Which animal, flower, sport would be the one most distant, in a sense most distasteful, to see associated with your activities, and what does that tell? Again: try to apply the same test, the same image or metaphor to the competition.

How to test the beauty of the objects you provide? The delight they give? What would qualify as elegance? What do they express? Do they get attention, and is it enough attention, the right kind of attention? Yes, one might claim that as well as competing for the buyer's money, you are competing for his attention. Here, the competition is no longer with your ordinary competitors producing products and services within the same market segments but rather other attention-capturing items and activities. Which are those, and how may we best grab that limited resource, attention? How might the attention span of the customer be broadened, if that is what is needed?

Are what you offer seen as producing positive experiences, challenging and empowering, not frustrating or irritating? How can you avoid boring or tiring your audience? How can your audience come to accept, adopt, internalize your products? How may they be brought to see them as statements or expressions of their own personalities?

What are the opposite attributes that pertain to the specific product category: Safety? Compact? Lightweight? Non-obtrusive? Non-invasive?

Established management techniques may, of course, be applied in the field of design as well. How would your products (services, etc.) compare with the competition's? What about clever benchmarking, that is, defining some important attributes, and then looking into some of the leading providers of products associated with precisely those attributes. Just like when someone looking to improve internal logistics would visit Federal Express to learn the secrets of a company entirely devoted to and dependent upon first class logistics.

From concepts such as management, benchmarking, and logistics to beauty and aesthetics — let us try making the step, with some help from a voluminous literature on those intertwined topics, though with a view, deep down, to how it all applies to design! Aesthetics is the philosophical notion of beauty while taste is a result of education and awareness of cultural values; thus, taste can be learned. Taste varies with class, cultural background, and education. Poor taste is usually seen as a product of ignorance. Aesthetic judgments can be very fine-grained and internally contradictory. But, may 'poor taste' constitute an important market category?

Beauty is associated with aesthetics, subject to a lengthy Wikipedia article, its extension and content reflecting the intricacies associated with coming to grips with this concept, nay, category. "Beauty is a characteristic of a person, place, object, or idea that provides a perceptual experience of pleasure, meaning, or satisfaction. Beauty is studied as part of aesthetics, sociology, social psychology, and culture." 'Enjoyment' comes out of plea-surable sensations, but for something to be 'beautiful' requires also that our capacities for reflective contemplation become engaged. Judgments on beauty are all at once sensory, emotional, and intellectual.

More Wikipedia: "A third major topic in the study of aesthetic judg-ments is how they are unified across art forms. We can call a person, a house, a symphony, a fragrance, and a mathematical proof beautiful. What characteristics do they share which give them that status?" Actually, what makes a painting beautiful is quite different from what makes music beautiful, implying, possibly, that each art form has its own language, its own values, for aesthetic judgments.

Wikipedia suggests that there is "quite a lack of words to express oneself accurately when making an aesthetic judg[e]ment." But in our attributes lists, we find a large number of words pertaining to degrees of beauty, attraction, or their opposites. This indicates the fact that precision is impossible, so that two completely different feelings experienced by two different people can be represented by the very same verbal expression. Wikipedia proposes that this is due to the inaccuracy of the English language, but Wittgenstein may be closer to the mark when he states that aesthetics is an arena for language games.

There is no lack of attempts to create categories and laws that would pertain to the domain of beauty.[1] It turns out to be much as with the philosophical concept of meaning, like when Augustine suggested that "I know what time is as long as you don't ask me to define it." Or simply that inherently subjective connotations defy such treatment to categories, hierarchies, laws, and what not.

When Goethe urges "Bilde, Künstler, rede nicht," he implies that beauty may defy words: it is its own explanation, self-referring. Wikipedia may be underrating the English language in the sense that it is not necessarily that particular language that lacks precision, but the object that defies such: beauty.

Puffer[2] contends that neither theoretical philosophy nor empirical enquiry has succeeded in formulating a consistent theory of beauty; the genesis and development of art forms is what is attainable.

"Thus it is in psychology that empirical aesthetics finds its last resort... Beauty is an excellence, a standard, a value. But value is in its nature teleological; is of the nature of purpose... A thing is not beautiful because it has value, [...] because it fulfills the end of Beauty..."

"So Beauty is again the pivot on which a system turns... The beautiful object possesses those qualities which bring the personality into a state of unity and self-completeness... at the highest possible point of tone, of functional efficiency, of enhanced life. Then a combination of favorable stimulation and repose would characterize the aesthetic feeling..."

"Now a beautiful object is first of all a unified object; the loss of the sense of personality is an integral part of the aesthetic experience; and we have seen how it is a necessary psychological effect of the unity of the object."

Similarly, philosopher John Armstrong[3] has concluded that the experience of beauty "consists in finding a spiritual value (truth, happiness, moral ideas) at home in a material setting (rhythm, line, shape, structure) and in such a way, that... the two seem inseparable."

In a way, these philosophers, artists, and authors may be seen to have grappled with the validity–reliability conflict outlined by Roger Martin.[4] Their explanations not of beauty itself so much as why explanations are hard or impossible to come by may serve as an underpinning to the suggestion that it is futile to look for one comprehensive and exhaustive design process.

In 2009, two magisterial works on beauty were published by Denis Dutton[5] and Roger Scruton.[6] They both decry the attacks on beauty in much of contemporary art, while they differ in their views of Art's origins. Dutton is, in the Darwin year, solidly Darwinian, Scruton is opposed to that idea, the idea that art evolved because it served a selection and survival purpose. Two artists, Vitaly Komar and Alexander Melamid, have carried out an extensive, systematic poll of the art preferences of people in ten different countries in Europe, Asia, Africa, and the Americas. The results demonstrate that what people appreciate of the picture contents in art is surprisingly uniform — worldwide.

Dutton's Darwinian argument is that literature, music, and much else have profound roots in human nature. Behind every artistic act lies the idea of a fitness test. Universally, we admire individual skill and virtuosity. "Skills and qualities of mind constitute eloquence, and the admiration of *eloquence* is solidly on the list of human universals."

Scruton[7]:

"Our need for beauty is not… a redundant addition to the list of human interests… not something that we could lack and still be fulfilled as people. It is a need arising from our metaphysical condition, as free individuals, seeking our place in a shared and objective world. We can wander through this world, alienated, resentful, full of suspicion and distrust. Or we can find our home here, coming to rest in harmony with others and with ourselves. The experience of beauty guides us along this second path: it tells us that we are at home in the world, that the world is already ordered in our perceptions as a place fit for the lives of beings like us… Hence the experience of beauty also points us beyond this world, to a 'kingdom of ends' in which our immortal longings and our desire for perfection are finally answered. As Plato and Kant both saw, therefore, the feeling for beauty is proximate to the religious frame of mind arising from a humble sense of harmony between the world around us and the needs within, and aspiring towards the highest unity with the transcendental... According to many philosophers and anthropologists... the experience of the sacred is a universal feature of the human condition, and therefore not easily avoided. For the most part our lives are organized by transitory purposes. But few of these purposes are memorable or moving to us."

Philosopher Dutton has also suggested seven universal signatures in human aesthetics, and we may contemplate how each of them applies to industrial design[8]:

1. Expertise or virtuosity. Technical artistic skills are cultivated, learned, recognized, and admired.

2. Non-utilitarian pleasure. People enjoy art for art's sake, and if it may in some respects be useful, the pleasure it gives is aside of any such practical or information consideration.

3. Style. Artistic objects and performances meet rules of composition that place them in a recognizable style though to a degree that varies greatly.

4. Criticism. People make a point of judging, appreciating, and interpreting works of art, and there is a language for that.

5. Imitation. With a few important exceptions like music and abstract painting, works of art represent or imitate experiences of the world.[9]

6. 'Special' focus. Art is set aside from ordinary life and made a dramatic focus of experience.

7. Imaginative experience. Artists as well as their audiences entertain hypothetical worlds in the theatre of the imagination.

So should we not talk about Beauty-Driven Design? It is worth reflecting on to what extent not designing for beauty first, creating or choosing the technology for it only next, might be a challenge, and the high road to success. Buckminster Fuller, for one, proposed that he was *not* designing for beauty but trying to solve problems, but if the outcome was not beautiful, then it was somehow not correct.

A man sat in the subway playing the violin. Nobody but sometimes a few kids noticed. It was Joshua Bell, one of the world's finest violinists playing incognito at a metro station, at the instigation of the Washington Post as part of a social experiment about perception, taste, and people's priorities. The themes were how we do perceive beauty in a thoroughly commonplace, not 'artistic', environment, and at an inappropriate hour. Do we really stop to appreciate it? Do we recognize talent in an unexpected context? So do we even appreciate beauty when not in a mood, or milieu, or context, attuned to appreciating it? Maybe that context must be *created*?

There are technologists who optimistically state that a solution is so superior that it will sell itself without any further ado. Most often, they are deceived. Likewise, someone who designed a beautiful artifact might claim its beauty to be so obvious as to be immediately recognized as such. Again, they may well be disappointed: sometimes the context must be right, sometimes taste has to be learnt, to be developed, to be taught. Possibly, fashion has something to do with this; some expressions of ephemeral culture co-evolve, catching people's minds in an intricate dance. There are some theorists (and perhaps practitioners) who claim that taste is an innate gift (or pain) that one is born with. But then it is most probably the same with creativity: we all have it, we may all develop it though to a larger or lesser degree. And it may be stifled or nurtured at school, at the workplace, in the family, and by society at large.

Santayana[10] emphasizes "the charm of symmetry." He contends that the "tensions of the eye" should be balanced so that we require bilateral symmetry.[11]

> "The necessity of vertical symmetry is not felt because the eyes and head do not so readily survey objects from top to bottom as from side to side... In other cases symmetry appeals to us through the charm of recognition and rhythm... when an anticipation is not met, the result is a shock. This shock... gives the effect of the picturesque; but when it comes with no compensation, it gives us the feeling of ugliness and imperfection..."

According to one theory, subjective perceptions of beauty correspond to compression of sensory input (Schmidhuber)[12]. Beauty equates with simplicity, in art, simplicity of a geometrical kind. The realization that an intricate sketch is the result of a limited number of geometrical transactions inspires fascination, makes the viewer appreciate the piece of art as one of beauty. Art is not paradoxical, it only seems so; it thrives on novelty but that novelty comes from discovering new aspects or applications of existing regularities and patterns. The opposite to simplicity is complexity, and Kolmogorov has suggested that a measure of complexity would be the length of the algorithm with which something is expressed; Kolmogorov's algorithm-based definition is but one of 42 mathematical measures for complexity (according to Seth Lloyd of the MIT; another is, just to take a random example, the Lempel-Ziv complexity). Thus, if beauty equates with regularity and can be reduced to being produced through the application of a few principles, it would also equate with simplicity, skirting complexity. The same argument, that relative simplicity equates with beauty, can be derived from Wolfram's *A New Kind of Science*, NKS.[13]

Wolfram:

"Everyday experience tends to make one think that it is difficult to get complex behavior, and that to do so requires complicated underlying rules. A crucial discovery in *A New Kind of Science* is that among programs this is not true — and that even some of the very simplest possible programs can produce behavior that in a fundamental sense is as complex as anything in our universe... Simple programs can yield behavior startlingly like what we see in nature... One bizarre possibility is that forms like those from rule 30 [one of Wolfram's 256 possible rules for creating two-dimensional black-and-white cellular automata] could have been created as art long ago but not be recognized now. For while it is easy to tell that a cave painting of an animal is a piece of purposeful art, dots carved into a rock in an approximate rule 30 pattern might not even be noticed as something of human origin."

"Many of the pictures in [A New Kind of Science] look strikingly similar to artistic designs of various styles. Probably this reflects not so much a similarity in underlying rules, but rather similarity in features that are most noticeable to the human visual system. Note that square grids of colored cells as in the cellular automata in this chapter [that is, in Wolfram's book] can be used quite directly as weaving patterns."

One may state that simplicity and efficiency are ever-present concerns in aesthetics. Physical and biological systems strive for economy, for least effort and energy consumption. If art and aesthetics — beauty — equate with simplicity and efficiency, beauty would serve as an important tool in and on which to thrive when natural resources dwindle. The evolutionary process of nature certainly results in intelligent design, elegant design, efficient design: maximum impact for minimum input. (See the section "Choosing simplicity — not so simple" later in this chapter for a further discussion of simplicity versus complexity.)

Horace Brock, who combines a mathematician's mind with the interests of an art collector and music expert, has recently published an intriguing theory of decorative arts, applying also to music.[14] The recurring items that others have fought with, symmetry, complexity, and simplicity, are also present in Brock's thesis. First of all, he states that beauty is not only subjective but *exists* — not just in the eye of the beholder. What constitutes a piece of art and of design are two factors working together dynamically, what he has termed a 'theme' and a 'transformation.' (One might suggest other denominations; having tried possible alternatives, I have concluded it best to stay with his chosen words; theme certainly works well with music.) The theme is an element that generates interest and tension like an S curve or converging lines; transformation stands for how that theme is worked upon to create an interesting, intriguing, and engaging piece of art. Again simplicity meets complexity; again our senses are engaged at several levels at once: 'an intricate sketch... the result of a limited number of geometrical transactions' as proposed previously.

Brock's eloquent and elegant reasoning is an important — perhaps *the* important — take-home message from this chapter. Whether this is a straightforward guide for performing a particular design task may be contingent upon that design project or, possibly, upon successfully translating the guideline into something adapted to the task at hand. Possibly, the other views on aesthetics referred to in this chapter can be of assistance, again depending upon the situation.

Verganti[15] has introduced the concept of product language, and language is a tool to establish order in reality, to reduce complexity. Design or product language, in this sense, is applied to convey meaning, but that meaning may be about the product itself (mundane and functional), about the purpose that it serves (light rather than lamp), about the user and her ambitions (a car or a house sending a message about its owner), about the experience it purports to offer (a car, a motor bike), or just what it is not (Philippe Starck's lemon 'squeezer'?). Themes and transformations would figure in that language, as categories to be filled with geometries, colors, translations, and what not applying to a particular corporate brand or product category.

Schmidhuber and Brock had a famous predecessor in George David Birkhoff, who was prominent in physics and mathematics but also had a keen interest in aesthetics, engaged in what makes a painting, piece of music, or poem pleasing. He, too, sought a formula, a mathematical expression, capturing beauty. The formula he came up with is alluringly simple: $M = O/C$, where M is the aesthetic measure or value, O aesthetic order, and C complexity. Thus, Birkhoff puts a high aesthetic value on orderliness, and a low one on complexity: beauty increases when complexity decreases. Most straightforward was the formula's application to polygons, but also less interesting. Birkhoff himself stated, "The 'complexity' of paintings is usually so considerable that they are analogous to ornamental patterns[16] whose constituent ornaments must be appreciated one by one. However, it is decidedly interesting to remark in this connection how a fine composition is always arranged so as to be easily comprehensible." Any work of art might, in his view, be analyzed with the assistance of lines given by the composition (a technique applied by Schmidhuber as well) and then the pattern of light and dark areas, too. "There should be a natural primary center of interest in the painting and also suitable secondary centers," Birkhoff explained.

Like with Birkhoff's polygons, designers working with, say, computer screens or web pages may restrict themselves to geometric measures. One such attempt gives five aesthetic measures: balance, equilibrium, symmetry, sequence, and order and complexity; whereas the same authors for a more complex graphic page suggest fourteen, several of them associated with degrees and forms of symmetry.

Umberto Eco has produced a book — in effect, two, the other dealing with the flip side, ugliness — on beauty: *History of Beauty*[17] (originally in Italian). Both books are rich with images of masterpieces of art and also replete with quotes from philosophers and artists who have expressed themselves on the subject, for the theme of ugliness often more obliquely than for those approaching the subject of beauty.

For the next few pages, we will follow Eco's recapitulation of the history of beauty through centuries, even millennia, of philosophical discourse.

The book on beauty begins

"'Beautiful' — together with 'graceful' and 'pretty,' or 'sublime,' 'marvellous,' 'superb' and similar expressions — is an adjective that we often employ to indicate something that we like. In this sense, its seems that what is beautiful is the same as what is good, and, in fact, in various historical periods there was a close link between the Beautiful and the Good."

Santayana[18] (whom Eco does not mention) says, "We know on excellent authority that beauty is truth... and the sensible manifestation of the good."

Eco's book on ugliness demonstrates that, perhaps, ugliness to an even larger extent has been equated with 'bad.' Eco continues,

"But if we judge on the basis of our everyday experience, we tend to define as good not only what we like, but also what we should like to have for ourselves [...] we realise that we talk of Beauty when we enjoy something for what it is, immaterial of whether we possess it or not. ... a well-made wedding cake... may strike us as beautiful, even though for health reasons or lack of appetite we do not desire it as a good to be acquired."

Beginning with Plato, there was uneasiness in allowing for the greatness of Socrates while recognizing how ugly he was. How to reconcile this conflict: recall that the very first quote from Eco equated beauty with the good, the ugly with something bad? The Greek word *techne*, the origin for technique as well as technology, was taken to mean that which is beautiful and functional; there was no conflict, beauty lay in the perfect function (*techne* would not apply to the charms of philosophers). Aquinas stated that beauty is the result not only of proportion/harmony (how the parts relate to the whole), brightness/radiance (the pleasure we may feel), or clarity but also of integrity (standing out from the background). He furthermore submitted that a crystal hammer that did not do the job of a hammer could never be considered beautiful even though made from a valuable material. Kant makes the distinction that a natural beauty is a *beautiful thing*; artistic beauty is a *beautiful representation* of a thing. We note that we design things, and, in this sense, Web pages are also such.

> "According to mythology, Zeus assigned an appropriate measure and a just limit to all beings: the government of the world thus coincides with a precious and measurable harmony, expressed in the four mottoes on the walls of the temple at Delphi: 'The most beautiful is the most just,' 'Observe the limit,' 'Shun hubris [arrogance]' and 'Nothing in excess.'"

Harmony, proportion, balance are, in fact, recurrent themes in the history onwards. The Greeks posited an antithesis between beauty and sensible perception; Eco contends that beauty is perceivable but not completely: a divide opens up between appearance and beauty. Furthermore, they submitted that there existed another antithesis between vision and sound, between idea on the one hand and passion and chaos on the other.

Among the many critics and philosophers quoted by Eco, Karl Rosenkranz figures prominently in both books: "Beauty reveals itself as the force that brings the rebellion of ugliness back under its control and dominion… Ugliness is relative, however, because it cannot find its measure in itself, but only in Beauty." Soon Rosenkranz arrives at the question of fashion:

> "The sphere of conventional Beauty, of fashion, is full of phenomena that, judged by the idea of Beauty, can only be defined as ugly, and never the less they pass temporarily as beautiful… but only because the *Zeitgeist* finds in such forms an adequate expression of its specific character and thus becomes accustomed to them. The *Zeitgeist* is more often in correspondence with fashion than anything else: here even ugliness may serve as a suitable means of expression."

The contention that beauty is profoundly subjective found an eloquent protagonist in Hume in the mid-18th century: "Beauty is no quality in things themselves: It exists merely in the mind which contemplates them; and each mind perceives a different beauty. One person may even perceive deformity… To seek the real beauty, or real deformity, is as fruitless an enquiry, as to pretend to ascertain the real sweet or real bitter." Kant claims that beauty pleases in an objective manner without being ascribable to any concept. Though in considering an object to be beautiful, "we hold that our judgment must have a universal value and that everyone must (or ought to) share our judgment. But, since the universality of judgments of taste do not require the existence of a concept to be conformed with, the universality of Beauty is subjective." The perception of beauty may well defy verbal description: 'Je Ne Sais Quoi' is an expression to claim the primacy in the eye of the beholder often associated with Rousseau but which Eco traces back to Italy and to Tasso.

In the very first sentence of Eco's book, he used the word 'sublime' as a synonym to beauty. Pseudo-Longinus, probably living in the first century AD, is the first to employ this expression to stand for the passions, the emotions evoked by a piece of art, or by nature for that matter. In the 18th century, it came to be associated with strong feelings, purification, and catharsis. But in that century, Burke suggested juxtaposing beauty and the sublime, the latter standing for vastness, ruggedness, negligence, solidity, flourishing in obscurity, or unleashing horror. For Burke, beauty was associated instead with smallness, smoothness, variety and gradual variation, delicacy and purity, grace and elegance. "In this series of characteristics it is hard to find a unifying idea," Eco submits. "For Kant the characteristics of the beautiful are: disinterested pleasure, universality without concept and regularity without law... we enjoy [the beautiful thing] as if it were the perfect embodiment of a rule, whereas it is a rule unto itself." Kant distinguishes between two types of sublime: the mathematical and the dynamic. The first is, e.g., the sight of the starry sky: we are induced to imagine more than we see. "A typical example of the dynamic Sublime is the sight of a **storm**. Here, what shakes our spirit is not the impression of infinite vastness, but of infinite power."

Eco continues:

"Today we are accustomed to talking about beautiful machines... But the idea of a beautiful machine is a fairly recent one... (...)... many poets have expressed their horror of machines. (...) ... machines appeared as quasi-human or quasi-animal, and it was in that 'quasi' that their monstrosity lay. These machines were useful, but disquieting: people made use of what they produced, but saw them as vaguely diabolical creations, and thus devoid of the gift of Beauty."

Watt tried to have the functionality of his first machine concealed under an exterior resembling a classical temple; the Eiffel Tower had to be made visually acceptable with classical inspired arches as pure ornament; cameras, typewriters, and textile machinery from the same epoch likewise feature such ornaments.

The tide turned completely with the futurists like Marinetti who stated that a racing car was more beautiful than Nike of Samotrace:

"We must prepare for the imminent and inevitable identification of man and machine, facilitating and perfecting an incessant exchange of intuitions, rhythm, instincts and metallic discipline..." "We must kneel on the tracks to divine speed. We must kneel before the rotating speed of a gyrocompass. 20,000 rpm, the maximum speed attained by man. We must steal from the stars the secret of their amazing, incomprehensible speed."

In his last few pages, Eco reaches our time and our society, offering as his last heading 'The Beauty of Consumption.' Positing the reactions of a visitor from the future, he sums up recent paragons of beauty on the cinema screen, interleaved with art genres of late like Pop Art. That visitor would have to "surrender before the orgy of tolerance, the total syncretism and the absolute and unstoppable polytheism of Beauty."

In his companion volume to *"History of Beauty," "On Ugliness,"* Eco[19] points to a void in the sense that ugliness never called forward any treatise of any length on that very subject, relegated to passing mentions in marginal works. Aquinas stated that beauty is the result not only of proportion, brightness, or clarity but also of integrity. There is a distinction between ugliness in itself and formal ugliness; a lack of equilibrium; and then the artistic representation of both. There can be a situational ugliness, a feeling of something as uncanny: balance and harmony upset. Exaggerations, at times, underline and add to beauty; sometimes, perhaps even more pronounced, it proceeds into caricature.

For Jung, "ugliness today is a sign and a symptom of great transformations to come... that will be appreciated tomorrow as great art could seem distasteful today." Certain musical intervals were perceived as dissonant and thus unpleasant. In the middle ages, this was the Devil (diabolus) in music; and later it was used to create and to design tension or instability.

"... ugliness is relative to the times and to cultures, what was unacceptable yesterday may be acceptable tomorrow, and what is perceived as ugly may contribute, in a suitable context, to the beauty of the whole. The fourth observation leads us to correct the relativist perspective: if the diabolus has always been used to create tension, then we have physiological reactions that have remained more or less unchanged through different periods and cultures. The diabolus has gradually been accepted not because it has become pleasant, but precisely because of that whiff of sulphur it has *never* lost."

The experience of beauty has two sides: euphoria and pain. Euphoria lifts us to a soaring feeling; pain resulting because we realize that beauty is like a rainbow, unattainable. We are prisoners in the less-than-perfect. To these two, we might add the sublime just described.

As so often recommended in creativity sessions: turn a concept upside down to see what happens. How about striving for what is awful, detesting or at least clumsy? Here too, and again, we can play the (partial) synonyms game: distasteful, hideous, disgraceful, unappealing, revolting, ugly…

It is even that beauty can be understood only with the existence of its opposite, that which is ugly or repugnant. Artist Jaymi Zents bases her work on the idea that beauty is characterized by contradictions.[20] She associates the opposites beautiful–ugly with other such pairs, like vitality–decay, or life–death (cf our lists below). She underlines that "the relationship of decay and beauty underlies most of my work in drawing, sculpture, and photography… The frailties of the flesh, its rashes, stains, scars, and the stress of everyday functions create narratives. The inherent conflicts of the body [and its] constant renewal, self-preservation versus the self-destruction of aging… the notion of the contaminated within the pristine." Beauty, a sense of beauty, develops, connecting meaningfully with that which conventionally is seen as the very opposite.

So ugliness can be seen as a necessary correlate for beauty to exist. All the more important for you to engage in understanding that side of the subject too! The beautiful is dependent on another opposite as well: it is dependent on what is in the beautiful object or piece of art, but also of what is absent. And then mastery equates with a capacity for limitations, for leaving some discovery and interpretation to the ones who enjoy the work: Lionardo spoke of sfumato, the diffuse, only hinted at, like a mysterious smile, fascinating and tempting generations of art interested.

Choosing simplicity — not so simple

Craigslist is a tremendously popular Web site — it gets more traffic than, for example, eBay or Amazon, companies that employ thousands and thousands of people. Craigslist has a staff of — thirty. They charge very modestly for ads, if at all: in many places, their service is for free. The site has almost 50 million unique visitors from the US each month. The design of the site is pure anti-design, as simple or poor as possible, like from the earliest days of the Internet. But the important point is that it works — simplicity is the key to a user experience of minimum time, problems, money spent. No learning curve! No series of touch-points with signing up, password, account confirmation, signing in...

In *Design-Inspired Innovation*, we emphasized elegance and simplicity as important qualities. They do not always carry the day, as we shall see, but there are many acolytes to the idea about the power of simplicity, like John Maeda[21] or the Philips corporation, with its Director for Simplicity.

For a while, *The Economist* magazine offered potential subscribers three options: on-line only for a certain price, print-only for almost the double, and online *and* print for the same. A lecturer designed a test with his students, and with these three options, 84 percent chose print and online, no one chose the phony print only, the rest or 16 percent chose online-only. When he removed the phony print-only alternative, 68 percent went for online only, and only 32 percent took the combination... just the *existence* of a third alternative changed the entire outcome.

Too many options, on the other hand, stifle decision-making. With a base of nearly 900,000 employees, given the option of choosing retirement programs, statistics are on a pretty solid footing. When employees were given two funds to select from, about 75 percent did enter the program; with ten, it was down to 70 percent; and with more than 30, it dropped to 60 percent. When a shampoo brand reduced its number of versions by almost half, sales rose by ten percent.

In a dairy shop, researchers set up a stand with a series of exotic quality jams, offering consumers to taste, given a one-dollar coupon as a bargain if they bought a jar. In one part of the study, six jam varieties were available, in another, 24 varieties. In both cases, all 24 varieties were available for purchase. The 24-variety array attracted more people to the test than the smaller one. There was a tremendous difference in buying behavior: 30 percent of those who met a small array of jams bought a jar; just three percent of those exposed to 24 varieties did so.

In another study, in a laboratory, college students had to evaluate an assortment of gourmet chocolates. The students were first given descriptions of and allowed to see the chocolates, to choose which they preferred. Only then did they actually taste and rate that particular chocolate. Finally, in another room, the students were offered a box of the chocolates, or, alternatively, cash for their participation. For one group, the chocolate array numbered six, for the other, 30. Students choosing from the small array were more satisfied with their tasting than those faced with the large array. Furthermore, they were four times as likely to choose chocolate instead of cash for their participation.

A Dutch thesis work[22] demonstrates that there is an inverse U-curve for people's preference for assortment size: small and too simple is bad, as is too large. People who are 'satisficers' (Nobel economist Herbert Simon's term) are content with a bit smaller assortments than 'maximisers.'

There is some controversy about the power of simplicity, however. First, there is the instance of something being so ugly that it flips over into being attractive in an odd sort of way. Then there is the distinction between allure at the point of sale and elegance in actual operation. An expensive investment good that seems to offer less features and less of user control cannot be prized higher than the feature-rich alternative, or can it? Don Norman,[23] not one to care for bad design or difficult-to-operate user interfaces, stresses that consumers actually *do* want complexity and features in much of what they consume. The controversy cannot be resolved completely but the point would rather be that an underlying complexity and desired features should be accessible through a user interface that is easy and elegant — the central quality is *capability*: "What people want is usable devices, which translates into understandable ones." And the way design is there to help manage complexity, making the unruly and confusing, instead, structured and understandable.

A striking example: in a Dutch town, Drachten, with 40,000 inhabitants, traffic engineer Hans Monderman had an upside-down idea. He eliminated all traffic signs but for those about speed limits and dangerous curves, installed a roundabout, and all busy intersections (the most busy saw 20,000 cars a day) became safer — and traffic faster. Direct, obvious simplicity created a 'shared space' between pedestrians, bikers, and car drivers: civility evolved. Right-of-way was now negotiated by direct human interaction, no longer by way of signs, which are often enough ignored. Sidewalks are at the street level but clearly denoted by texture and color.

Everything elegant is simple; the opposite is not necessarily true — not everything simple *is* elegant. There are at least two kinds of simplicity. The elegant one is powerful: maximum effect with least effort. Composers and theatre actors use silence to create dramatic effects;[24] powerful pieces of art depend on what is purposefully left out. When Toyota created, in effect designed, a new youth brand, Scion, they abstained from ordinary advertising and drastically reduced the number of vehicle features so that ad-averse Generation Y buyers could make personal statements, customizing their cars with trendy accessories.

Attributes

As advertised early on in this chapter, here follows a series of lists of attributes. The idea is to give you the raw verbal material to test and try how they apply to the products and services that you are providing. The lists are long but not exhaustive; the English language, and not just that language, is rich. There are some attributes that can be seen as opposing, and as you browse further, there is a brief discussion of some of the features highlighted through the lists and their inherent qualities.

Positive attributes (our qualifications may well be discussed, though)

accommodating	cozy	fashionable
aesthetic	crafty	fine
agreeable	creative	flabbergasting
alluring	cute	flourishing
amorous	daunting	flowerful
appealing	dear	forceful
appropriate	delicate	formidable
attractive	delicious	fruitful
awe-inspiring	delightful	fruity
awesome	desirable	fulfilling
balanced	distinguished	fun
becoming	driven	genial
brilliant	durable	glad
captivating	dynamic	golden
cheerful	easy	good
classy	elegant	gorgeous
clean	empowering	graceful
clear	enthusing	gracious
clubby	essential	grand
cocky	ethereal	great
comely	expansive	gregarious
communicative	expressive	handsome
congenial	exquisite	happy
congruent	extraordinary	harmonious
convenient	fanciful	healthy
convincing	fantastic	hearty
cool	fascinating	homely

honest
honorable
ideal
imaginative
important
impressive
inspiring
integrity
intelligent
inviting
jolly
joyful
lively
lovable
lucky
masterly
mighty
mind-blowing
open
outgoing
peaceful
perfect
pleasant
pleasing
pleasurable

plentiful
positive
potent
powerful
pretty
refreshing
resilient
resonating
respectable
satisfactory
satisfying
sensational
shapely
simple
skilled
sober
solid
spacious
splendid
stable
startling
sterling
stylish
sublime
subtle

superb
superior
sympathetic
tasteful
thrilling
true
truthful
unbelievable
unforgettable
upbeat
uplifting
useful
valid
valuable
venturesome
vigorous
virtuous
wholesome
wired
witty
wonderful
wondrous
worthy

Negative attributes

abhorrent
abominable
abrasive
antipathetic
aping

appalling
ashen
atrocious
average
awful

awkward
bad
baleful
barbarian
baroque

base
belittling
besmirched
bigre
boasting
boisterous
brusque
bungling
bureaucratic
cannibalistic
carnivorous
cheap
conspiring
contemptible
copying
corrupt
crude
cruel
cumbersome
deceiting
deceiving
degenerate
depraved
despicable
detestable
devastating
deviating
dilettantish
dirty
disastrous
disgusting
distasteful
distasting
doleful
doubtful
drab
dreadful
dreary
drunken
dubious
dysfunctional
dystopian
earthen
eerie
egregious
empty
evil
execrable
expensive
extravagant
faceless
false
freakish
freaky
frivolous
fitful
forbidding
forced
foul
frightful
fumbling
fuzzy
ghastly
glib
glum
goofy
gory
grim
grisly
gross
grotesque
gruesome
grumpy
gullible
hateful
heinous
hellish
hideous
hilarious
homely
horrible
idiotic
imitating
impotent
inconvenient
infamous
infernal
insipid
intrusive
lame
laughable
leaden
loaded
loathsome
lousy
lurid
mean
meddlesome
meek
monotonous
monstrous
moody

nasty
nefarious
negative
obnoxious
odious
offensive
ordinary
outmoded
outrageous
painful
perverse
plain
plump
pointless
powerless
preposterous
problematic
putrid
rank
raucous
raw
repellent
repulsive
revolting
ridiculous
rigid
rotten
sad
scandalous
scary
schizoid

shameful
shitty
shocking
simplistic
slattern
solipsistic
sorrowful
sorry
stale
stony
stupid
sucking
superfluous
tame
terrible
toothless
torpid
trashy
treacherous
turgid
ugly
unaccommodating
unaesthetic
unbecoming
unclear
unconvincing
uncouth
uncreative
undistinguished
unfashionable
ungodly

ungraceful
ungracious
unhappy
unharmonious
unimaginative
unimpressive
uninspiring
unlovable
unsatisfactory
unsavory
unshapely
unskilled
unstable
untrue
vacuous
vague
vegetating
vicious
vile
villainous
vulgar
weak
weary
weird
wholesome
wicked
wooden
wretched
wrong

Attributes neither evidently negative nor positive

abstemious
aggressive
allowing
androgenic
antediluvian
antique
archaic
architectural
astenic
astonishing
auspicious
besotted
bionic
bizarre
bumpy
chubby
classical
closed
clueless
cold
colorful
colors: black,
   white, green,
   red, blue, yellow...
comic
compact
conspicuous
controlled
curvaceous
cussy
daunting
delicate
demure

derivating
derivative
devastating
direct
disturbing
down to earth
driven
dry
dynamic
easy
edgy
emotional
enforcing
ephemeral
expensive
extrovert
factual
fast
fearsome
finicky
flamboyant
flexible
flowering
fluid
foldable
formal
garish
gaudy
geometric
glacial
gregarious
heavy
hellish

hilarious
humble
humourous
imaginary
inauspicious
inconspicuous
indirect
intellectual
introvert
irrational
lean
lenient
light
light-weight
lively
lofty
loud
luggable
lukewarm
melancholy
mellow
middle-of-the-road
mobile
mocking
mollifying
moody
muckraking
multi-faceted
mundane
neutral
non-invasive
non-obtrusive
oblique

odd	retro	stony
onerous	rhythmical	subdued
ordered	robust	surprising
painless	safe	symmetrical
pale	satanic	technical
pious	scarce	tolerable
plentiful	self-contained	tough
polite	serene	transparent
politic	serious	two-headed
portable	shy	ubiquitous
practical	simple	unassuming
precise	slattern	unbalanced
prolific	slow	unfashionable
proprietary	solitary	unreal
prosthetic	soothing	utilitarian
protean	spartan	utopian
proverbial	sprightly	vague
prudent	sprighty	vegetarian
pure	spurious	virtual
quiet	staid	warm
relaxed	stolid	youthful

Clearly, a list of adjectives and attributes that are positive does not necessarily contain exact or even proximate synonyms. 'Clear' does not equate with 'clean,' 'becoming' certainly not with 'convenient.' Some words may display affinity and thus be grouped together; what about awesome, awe-inspiring, gorgeous, striking, handsome, graceful, harmonious, joyful, and more?

Some such groups may be seen as different grades of the same quality. Would not plain–meek–wooden–vacuous–empty–toothless–stale–pointless–faceless–doubtful–lame–unlovable–dubious–sucking–unbecoming–awkward–trashy–ugly–dreary–bad–awful–

dreadful–frightful–gruesome–terrible–hateful–horrible–abominable (by the way, would you agree with that order? Difficult to distinguish, and context-dependent, *n'est-ce pas*?) represent some kind of hierarchy, in at least some contexts?

Context is a key word. The context might be the type of design: cars differ from fashion clothes and chinaware. It might also pertain to what the design is aiming at, conveying an experience or producing a sustainable life cycle.

Depending on the product (or service) and also the kind of niche or particular attraction one wants to permeate, those attributes may have to be shifted around. Perhaps 'plain,' 'dubious,' or even 'frightful' might be impressions that are aimed for, attractive enough as markers of distinctions to be driven home. 'Clean' may, in a certain context, be interpreted as something not in opposition to 'dirty,' 'fashionable'; or 'sensational' may not be at all that positive, 'ethereal' or 'sublime' may give the wrong impression. 'Powerful' and 'forceful' may convey the wrong messages for products emphasizing brittleness, subtlety, and precision. On the list of negatives, 'ashen', 'barbarian', boisterous, or 'baroque' may well convey exactly what is fitting, and 'cheap' is certainly a positive value in many contexts, though then perhaps phrased as 'value for money.' 'Deceiving' is again a word open to interpretation, and 'undistinguished' may be a distinction for phenomena best performing when not being seen, not being in the least intrusive. From the list of 'neutral' attributes, many may be situated in a context where they are either entirely positive or entirely negative. Just look at aggressive, antique, archaic, astonishing, auspicious, bizarre, closed, cold, colorful, comic, and conspicuous…

One may establish opposites through the simple addition of a prefix such as dis-, im- or in- (there were lots of un- in the list for negatives). Some such are found in our tables and you may well create several more. Here are some attributes created with prefixes; with or without these, quite a few have already been listed:

(dis)tasteful	(un)clean	(un)imaginative
(im)potent	(un)clear	(un)impressive
(in)auspicious	(un)conspicuous	(un)inspiring
(in)convenient	(un)creative	(un)lovable
(in)direct	(un)distinguished	(un)pleasant
(un)accomodating	(un)fashionable	(un)shapely
(un)aesthetic	(un)graceful	(un)skilled
(un)appealing	(un)gracious	(un)true
(un)becoming	(un)harmonious	(un)wholesome

Some attributes are the opposites of each other — polar — without necessarily (depending on context) carrying positive or negative connotations, cf. the prefixes un-, in-, and others above:

slow–fast	knowledge–ignorance	open–closed
usual–unusual		height–depth
aggressive–friendly	mortal–immortal	wide–narrow
mature–youthful	profound–superficial	deep–shallow
masculine–feminine	human–divine	tall–short
old–new	human–animal	soft–hard
modern–traditional	heaven–hell	smooth–rough
sober–cheerful	earth–sky/space	particular–general
serious–playful	all–nothing	natural–artificial
passive–active	moral–immoral	outgoing–taciturn
solid–frail	right–wrong	stable–change
matter–spirit	good–evil	hierarchical–egalitarian
body–mind	virtue–vice	
conscious–unconscious	inner–outer	freedom–slavery
same–different	night–day	guilt–innocence
reason–feeling	dark–light	comedy–tragedy
	summer–winter	visible–invisible

harmony–disharmony	amicable–aloof	past–present–future
intriguing–ordinary	empty–full	lower–middle–upper
simplicity–complexity	life–death	class

Components of user experience: useful–desirable–valuable–accessible–findable–credible–usable–reliable–sustainable…

Endnotes

[1]For a discussion, see, for example, http://plato.stanford.edu/entries/aesthetic-judgment/.

[2]Puffer, Ethel D, 2003. *The Psychology of Beauty*. Project Gutenberg Literary Archive Foundation Etext #3751, Oxford, MS.

[3]Armstrong, John, 2004. *The Secret Power of Beauty*. Allen Lane, London.

[4]Martin, Roger, 2009. *Design Thinking: The Next Competitive Advantage*. Harvard Business School Press, Boston, MA.

[5]Dutton, Denis, 2009. *The Art Instinct*. Bloomsbury Press, NY, NY.

[6]Scruton, Roger, 2009. *Beauty*. Oxford University Press, Oxford.

[7]In *Axess Magazine* 2008:7 (Oct).

[8]http://www.denisdutton.com/universals.htm.

[9]Not everybody would agree…

[10]http://books.google.se/books?id=EOboyaZ6iSUC&dq=Santayana+beauty&printsec=frontcover&source=bl&ots=lGjpUjq4UW&sig=96upwS-o3R3QQzYDjLjjy5ESfwc&hl=sv&ei=tKs8Ss7MA4KRjAfVtOgY&sa=X&oi=book_result&ct=result&resnum=3.

[11]Symmetry is one dimension, color another, on which there is excellent information found at http://www.colormatters.com/entercolormatters.html.

[12]Schmidhuber, Jürgen, 2007. *Simple Algorithmic Principles of Discovery, Subjective Beauty Selective Attention, Curiosity & Creativity, Joint Invited Lecture* for *Algorithmic Learning Theory* (ALT2007) and *Discovery Science* (DS2007), Sendai, Japan.

[13]Wolfram, Stephen, 2002. *A New Kind of Science*. Wolfram Media, Champaign, IL.

[14]Brock, Horace Wood, 2009. *The Truth about Beauty*. An excerpt from the catalogue *Splendor and Elegance* published by MFA Publications for the exhibition "Splendor and Elegance: European Decorative Arts and Drawings from the Horace Wood Brock Collection," Museum of Fine Arts, Boston.

[15]In, e.g., Utterback et al. 2006. *Design-Inspired Innovation*. World Scientific Publishing, Singapore.

[16]Wolfram refers to such patterns, cf. Chapter 10.

[17]Eco, Umberto, 2004. *History of Beauty*. Rizzoli, NY. As mentioned, the works are full of quotes, put in a context, and Eco is actually credited with being editor.

[18]Op cit.

[19]Eco, Umberto, 2007. *On Ugliness*. Rizzoli, NY.

[20]http://rattengift.livejournal.com/5888.html.

[21]Maeda, John, 2006. *The Laws of Simplicity*. The MIT Press, Cambridge, MA.

[22]http://scriptie.jeroenvisser.com/.

[23]http://www.jnd.org/dn.mss/simplicity_is_highly.html.

[24]Just as an example: in his rendering of 'The Little Drummer Boy' (1962), piano player Henri Plessier demonstrates this brilliantly.

Chapter 6

MULTI-DIMENSIONAL CREATIVITY

Designing is closely associated with creativity. Design firms often claim to rely upon various, sometimes proprietary, creative or ideation methods, brainstorming being one method that figures prominently.

We have already encountered a number of techniques for enhancing creativity and generating new ideas, or variations of existing ones. When later, in Chapter 6, more in detail discussing sketching, we will underline that sketches should preferably be rough early on, holding more creative potential, and numerous, so as to inspire thinking in alternatives; likewise for early, frequent, speedy prototyping. When 'broadening the scope' in the next chapter, we will open the vistas for a wider input and, again, for the generation of ever more viewpoints and alternatives. We also saw, e.g., when in the introduction discussing the effects of native language on thinking, that diversity in a design team is highly preferable.

There is a healthy body of research and a great number of books on the subject of creativity, including many recommendations for generating wonderful ideas — sometimes without the necessary corollary that ideas are not born wonderful; no, a seemingly brilliant idea is never enough since it has to be realized to be of any utility; the proof is only in the pudding. This chapter will attempt to sum up some important dimensions of idea generation more closely associated with design work.

Creativity methods

Here, as already indicated, brainstorming is bound to be brought up. Most often, established wisdom does not take into account the circumstances when applied specifically to design, where it is frequently performed by the same team regularly. This is something quite different from the 'standard model' for brainstorming, with an ad hoc, temporary assembly of contributors. One rule that is important for ad hoc brainstorming is 'no criticism', but this may be substituted by 'constructive criticism' within a design team. A more sophisticated further development of brainstorming is Synectics, which, among other dimensions, focuses on constraints, sometimes lifting such. This is an avenue that might be pursued without necessarily keeping to the particular Synectics rigor. Among several more formalized approaches, we find, e.g., TRIZ*, with its origin in an analysis of patented solutions, that proposes a number of fundamental laws in science and technology and which can be built upon to generate alternatives systematically. Brainstorming may be supported by a set of alternative generators (make larger, smaller, turn upside down, combine, eliminate, etc.); whereas one of several formal methods of going through possible combinations when the design object is one that can be described in a number of dimensions is morphological analysis.

---

*Somewhat awkwardly, this is the Latin letter abbreviation of the original Russian: the theory for inventors' problem-solving.

Storytelling

Storyboards are often good tools for capturing essential elements in an experience, as might be more full-blown storytelling. Stories offer stimulating alternatives to other ways of viewing the problem or its context; they also make for a better understanding of the touch-points where a design and its user meet. And they have a prototypical, archetypal power. A particular type of storytelling is establishing several personas, those vivid snapshots of end users that we have tried to envisage when discussing how to approach the user.

Going into a parallel world

Synectics relies also on tools such as role-playing and analogy. Role-playing, theater, and body language belong to media offering additional possibilities for sneaking up on problems in new ways, possibly provoking original solutions. Biomimicry implies borrowing from nature, from its evolutionary process geared at optimization and economy; scene-mimicry would be like benchmarking: going into a situation from another field but with important similarities. Projecting into the future, thinking about how the designed object or service will play out in X years' time or preparing a handful of different scenarios offer yet other opportunities. Visiting the future is one way of creating psychological distance, but there are more, and they, too, have proven to be potent at times.

Diversity

We have underscored the power of widening the scope, returning at length to it in the next chapter; it is something that implies adding inputs from a wider variety of sources. The general recommendation of diversity may be interpreted in manifold ways. In the area of cultures and languages, we have seen the role of language on thought. We have explored age groups and individual ways of thinking (the Myers-Briggs classification may be debated, not the fact that there are several different ways of applying logic — several types of logic, as it were: concrete versus formal thinkers to mention one distinction). And thought styles, as well as various levels of expertise — expert chefs versus children in the kitchen or for designing kitchen tools — and what about designing for animals? Metaphors embody language, and their connotations may become very physical, or physiological, resulting in important caveats and opportunities. Looking for inspiration in new scientific findings and new technologies, not least materials, contributes to diversity in thinking and input.

Creativity enhanced

Creativity methods are sometimes divided into systematic and inspirational, brainstorming an example of the latter. Morphological analysis, a systematic one, attribute listing another one, as is the cross impact-matrix, which means going through all possible combinations of the elements, functions, or components of a system. Though seemingly mechanical, the number of combinations may be so large that intuition is stumped and an overview is too shallow for a fine-grained matrix. Let us take an old celluloid film-based camera as an example. The components are body, viewfinder, lens, film, and film winding mechanism:

Body	*Viewfinder*	*Lens*	*Film*	*Winding*
One piece	See-through	Interchangeable	120 roll	Wheel
Bellows	SLR	Fixed	35 mm	Lever
	Twin reflex	Fixed zoom		Motor

This table is not exhaustive (a 120-roll film may give pictures that are $4/5 \times 6$, 6×6, or 6×9 cm; 35 mm, 24×18, 24×24, or 24×36 mm) but serves our purpose of illustrating. The number of combinations is $2 \times 3 \times 3 \times 2 \times 3$ or 108 in total, and morphological analysis means going through them all. Twin reflex plus bellows would be thought decidedly original but, as a matter of fact, twin reflexes with bellows have indeed been produced.

Another method relying upon systemization is to construct an abstraction ladder, where order is established through categorization. Tables would be a category encompassing dining tables, bed tables, etc.; the same for chairs which, on the next level, would name different chairs so an office chair might be distinguished by having wheels or not, be reclining or not, etc., and then, on yet another level, marquee such as Steelcase Criterion, IKEA Markus, etc. If style is important, categories may include, for example, rococo, empire, baroque, etc. As with other types of analysis, the effort inspires new insights and unusual combinations. To move up the ladder and increase abstraction, one may ask *why?* or *why else?*, to go down and become more concrete, the questions would be *how?* or *how else?* Several more methods for systematizing and tweaking the system exist, such as attribute listing; the series of questions like: doing the job differently?, doing an adjacent job?, avoid the job completely?, doing it under different circumstances?, complementing it?; their applicability always contingent upon problem and situation.

TRIZ is based upon an exhaustive set of scientific and technical principles, compiled in a 39 × 39 contradiction matrix (for 39 engineering parameters) and 40 principles for problem solving. There is a whole body of teachings, including an entire language or terminology, such as distinguishing between administrative (having to do with needs versus abilities), technical (trade-offs inducing), and physical (inherent) contradictions. A specific problem is abstracted to the level of a general one, where a general solution can be sought, then resulting in a specific solution. This rich source of knowledge also contains, among other results generated, trajectory laws for how technology evolves.[1] Just examining patents aiming at delivering a particular function closely can be genuinely illuminating and inspiring (though it might seem discouraging at first: all these ingenuous solution and… the result?).

Brainstorming has been tried, tested, and developed by, among others, researchers and practitioners associated with the Creative Problem Solving Institute with close links to the New York State University in Buffalo. A basic principle developed there is that what is perceived as a problem is often not the fundamental one; solution-seeking starts with a 'mess.' To arrive at the problem to be solved, questions like 'in what ways may we...' should be posed. Reframe the problem!

This is something that we have touched upon previously trying to find out what is the real competition, for money or for time or, perhaps, other activities or allure, rather than the seemingly obvious: another firm in direct competition. Looking for a hole rather than a drill is another example belonging to this category. Industries and companies often operate on formulas that are taken for granted; the classic example is when Toyota figured how to compete with General Motors or Ford with just a tiny production without much of regular economies of scale. Another example is when the Swedish latecomer in the field of hard metals, Sandvik, emerged as the world leader because it changed the established business logic and concentrated on *teaching* customers the benefits of hard metal tools compared to steel. In Europe, Ryanair has up-ended passenger air transport. They are shunning regular airports in favor of also-rans that want to establish themselves and their larger regional constituency; they are not even no-frills — they have the passenger do much of the job; 25 minutes for the on-the-ground turnover; no connecting flights, pure point-to-point. An important conclusion is that the very problem that constitutes the 'mess' may well generate a whole ensemble of sub-problems with a potential for addressing the issue initially perceived.

A thorough study[2] of industrial designers in work sessions, and also, as a comparison, of improvising musicians, has gained deeper knowledge on how 'brainstorming' or, perhaps, rather 'idea development' can take place. Designers can both offer criticism and hold on to pet ideas as well as they can playfully decide to leave such a pet idea on the side for a while. What makes for constructive criticism is that a design team feels at ease since there is trust built up over time and several projects; there is no prestige or status at stake. When improvising musicians are inspired to achieve creative harmony, relying upon ambiguity and paradox, they 'rest in the groove' and experience a happy feeling of everybody contributing; the same can be achieved in designers' sessions and is an experience of *flow*. In improvised music, 'errors,' mistaken departures from the script, are taken as inspirational challenges offering the other musicians triggers for further creative contributions, generating a harmonious whole of profound novelty. Design session participants must, of course, listen carefully to each other — as musicians certainly do! — but since diversity is important, the presence of different thought styles should be respected and built upon, with no attempts made to harmonize or merge these styles.

The huge North American motel chain Motel 6 had no reputation for good design, or even for any design at all. Executives realized that there might be value in a profound redesign effort, and applied something akin to a bench marking or bench-learning project. One of the design firms recruited had been responsible for airplane cabins for Virgin Atlantic and cruise ship cabins for a Norwegian company. The results are striking: carpets have been ripped up, wood-effect flooring gives a spacious look, ambient lights substitute ordinary lamps. Constraints, not least about cost, resulted in smart design solutions: one unit contains flat-screen TV, closet space, and lighting system. Luggage is stored under the new platform beds.

Verbal language, indeed, tends to make for the overshadowing of thoughts — particularly where verbal descriptions are lacking. Ways around this include mute 'brainstorming' relying upon, for example, sketching or body language. It is also important to give freedom for silent or tacit knowledge. One way of doing so is to use the Dialog method were participants are invited to comment on and discuss challenging, rich works from art and literature seemingly with only tenuous relations to the core of the task;[3] the discussion should be contentious, meeting the challenges of the text.[4] Esther Henwood[5] has used literature as the point of departure for design, asking famous designers to take one book each for which, then, to design an object, and her book cover features, among several objects, Sapper's whistling kettle. Most products in the book are pieces of furniture, most books novels, but one that is definitely not a novel is Prigogine and Stenger's "La nouvelle alliance" about dissipative processes.

The more critique is part of the procedure, the more productive it becomes: skills in critiquing and in accepting critique grow. The task is to arrive at a better understanding — jointly. As mentioned, trust is an essential basis for critique to be effective, that is, respect and objective critique, not personal or competitive. It must be clearly understood that the person whose idea is the object for the discussion has provided an important contribution well worth discussing. Timing is essential; never create the feeling of an ambush, and beware of making recommendations — rather, inspire the one whose idea is discussed to come up with what s/he thinks should be alternatives or improvements. Think 'constructive contrarian' and have several people, including the idea champion, take turns playing that role. De Bono's *Six Thinking Hats* must be mentioned in this context.[6] They legitimize criticism by stating that the hat, now activated or carried, is the black one, offering critique. Likewise, for the other hats, the white aims to go into facts and information, the yellow sees the sunny, positive side, the green encourages ideation, the red delves into feeling and emotion, and the blue takes in the larger picture.

Yet another variation is reverse brainstorming. Here, instead of brainstorming *for* something, brainstorm for its direct opposite. So instead of looking for ways to turn a project into a success, the task is 'how can we make this project a total failure?' Try it, and hostility and negative feelings are turned into a somewhat weird but productive kind of creativity — it is easier and more fun to think about failures... Then, the list of all factors that would ensure failure can be prioritized and each of them subject to scrutiny to see how they can be avoided. Or try, for a while, to generate profoundly bad ideas — vaunt some such, and enjoy them.

Before the verbal stage, before logic and linguistics, creative thinking is channeled into emotions, intuitions, images, and bodily feelings. Verbal language is mute for music, whereas the piano key is potent. We may try seeing and experiencing with our bodies, seeing with touch. Smell is another trigger, and is often very powerful.

Some of the tools for fostering a process of design inspiration work to support memory, helping people to register and reflect on daily activities: a diary, or some electronic variation of it; perhaps a frustration diary logging everyday irritations, small and large; alternatively, the design team may establish a purported log for 'a day in the life of the person we design for.' There should also be inspiration from and assistance for dreaming, and for feeling. In these tasks, images may serve as triggers: people are asked to use collections of images as tools for expression. Results may be products that seduce; trivial mundane objects that normally are kept out of sight may instead be placed in the spotlight because of the delight they inspire.

Sometimes it is a question of seeing, like great artists do: seeing that which is blindingly obvious. Sometimes it is about discovering what is happening at the margins; sometimes it helps being naïve, being a child, seeing with a beginner's mind. The warm-up for a brainstorming session is important for setting the stage for taking the fresh look, and for being playful. Philips has developed 'Spark,' a fun game to further creativity in brainstorming and workshops, with good results.[7]

Art is not just what it is, but also what it is not, so important questions pertain to what should be left out, discarded, withheld, secreted away, hidden — there are differences between the meanings of these four concepts. The creative stimulus from sketching might be seen as giving opportunities for confrontations between ways of seeing and understanding, with representations provocative in themselves (sketch or mockup versus idea and the mind's eye), in types of representations; for example, mathematical versus verbal, and in dimensions of a task. One such a dimension might be discovery, through connections and associations, sliding from one idea to another, and a dialog between discovery and understanding.

Constraints are part of reality but fiddling with them is, as Synectics demonstrates, a way to ideation, to tweaking the problem. Synectics sometimes lifts constraints, but imposing new constraints or changing them in interesting ways may gain likewise interesting results. What about designing a car as a street lamp; what about drastically reducing the number of parts in the object (a car features some 30,000–50,000); what about applying lean thinking; what about thinking of photos of the invisible, from X-rays or microscopes, or simply printing it out in different or false colors, or starkly accentuated contrasts? With resources scarce, designs need to be better; bloated software demonstrates the opposite. Martin[8] underlines that poor design briefs are not necessarily those with many constraints but rather the ones that are posited so as to eliminate any opportunities for surprise and discovery.

In *Design-Inspired Innovation*, we praised Google for its minimalist design — just those 28 words — arousing pure delight and underlining the importance of simplicity. Thus, it is interesting to see Google losing part of the information market for failing to apply the same lesson there, handing the majority of a particular niche to rival Yahoo! When it comes to financial information and news, Yahoo! carries 17.5 times as much traffic as Google. So while Yahoo! is found at the top of the traffic in this category, Google is relegated to a dismal 17th place. Simplicity versus more data and charts, and the outcome is... that over the year July 2008–July 2009, Yahoo! enjoyed a 12 percent increase in traffic, Google just 3 percent.

A strong component in de Bono's lateral thinking toolbox is forced association: generate a random word from an encyclopedia, and try to force a relationship between it and the problem or design object. Or be inspired by all those words that are threatened by extinction, subject to a dedicated Web-based savings effort.[9] The method can be enlarged: collect toys, materials, components, images, videos that spur fantasy and evoke dreams, anything provoking flights into the unknown.

A way to seeing something from another perspective is to simply dissect, or tear down, an existing product. A number of questions arise related to design choices made, material, techniques, features included and excluded. As ever so often, the devil may be in the details.

Storytelling

Alberto Manguel[10] once said

> "… stories can assist us… they can heal us, illuminate us, and show us the way. Above all, they can remind us of our condition, break through the superficial appearance of things, and make us aware of the underlying currents and depths. [...] In ancient Anglo-Saxon, the word for poet was *maker* [...]"

It is often suggested that a company, a brand, should tell a story — a design should do likewise, since it is instrumental in creating the brand image. Often, the story is abbreviated into a brief catchy slogan: Just do it; Connecting people; Taking you forward; Sheer driving pleasure. Ideally, this phrase, together with the brand and its history, evokes a story that is attractive and that corresponds to the values and qualities that create the mutual understanding and expectations between producer-designer and customer-user. So one point of departure might be the story to be told, to be conveyed through the design(s).

The next several paragraphs rely heavily upon the magisterial (not undisputed) work of Christopher Booker who, after perusing a vast amount of literature in the form of novels and dramas, but also myths and sacred texts and even movie plots, has concluded that there are, basically, seven basic plots.[11] These are, in an all too brief rendering of a 700 pages book:

- overcoming the monster
- rags to riches

- the Quest
- voyage and return
- comedy
- tragedy
- rebirth

The Monster is a superhuman evil force, holding captive a beautiful princess or a rich treasure, and the hero sets out to fight the Monster. The odds are bad and the hero is almost destroyed, when in the last minute there is a dramatic turning point. As with our attributes for beauty or its opposite, a number of Monster attributes apply: terrible, misshapen, bloodthirsty, ravening, treacherous, vicious, vile, hellish, demonic.

The story develops in five stages: anticipation where curiosity is awakened; dream stage when preparations are made orderly; frustration stage when the Monster reveals all its, well, monstrosity; nightmare stage when all seems lost; and finally the thrilling escape and death of the Monster.

Likewise, 'rags to riches' plays out in five acts: initial wretchedness and 'call'; out in the world, early success; central crisis; final ordeal; final completion and success.

The Quest also starts with a 'call,' and the hero is supported by a number of fellow companions as well as helpers. The journey is an obstacle-race where again monsters, temptations, deadly dilemmas, and the underworld beckons. Here, the five steps are: call; journey; arrival and frustration; final ordeals; and winning the goal.

The voyage and return is a journey quite different from the Quest, sometimes social rather than physical. The five acts are: waking up to something deeply unfamiliar and unexpected; initial fascination with new discoveries; difficulty and oppression; downright nightmare; and escape and return.

Comedy is something else; here, Booker suggests the analogy of the jigsaw puzzle. The key is the transition between two states, and one or two of the earlier plots may be drawn upon. There is a small world where confusion is arising, possibly through the appearance of a dark shadow. Finally, the shadow and the confusion are eliminated and all is happiness. A whole network of relationships is involved, initially knotted because something central has gone wrong. A fundamental feature is concealed, the persons involved blind to it — until the happy resolution.

Tragedy is a story of inversion: falling for a forbidden temptation, trying to achieve a worthy goal but in a forbidden way, the hero is at first immensely successful, then things start getting off the tracks, and eventually the hero is destructed. Again five stages may be suggested: anticipation and desire — dream and fulfillment — foundations start trembling — things go seriously out of control — outright destruction or death wish. A number of personalities might be involved: the good old man, the rival or 'shadow,' the innocent young girl, and the temptress.

Rebirth, finally, evolves somewhat in parallel to the Monster story: a heroine or hero falls under the thrall of a dark force — the darkness seems to have receded — but darkness strikes and seems to have won; this imprisonment or state of living dead seems to have been established — until miraculous redemption. Darkness has a series of well-known attributes such as coldness, hardness, decay, torment, despair, as has redemption: softness, liberation, growth, hope, and love.

While noting that the last two centuries have seen deviations from the seven plots, though they are still somehow very present, Booker concedes that there are other plots as well, like creation myths and mystery stories. Likewise, there are quite a few novels and films that are composed — designed — to weave several, possibly all seven, plots into them. They may also be decomposed into smaller recurring themes like the amazing escape from death, the conquest over the dark shadow, the union of hero and heroine following upon separation. Booker underlines that these figures, as well as the basic plots, exist in all cultures and, thus, reflect something profound in the human condition, more than touching upon Jungian archetypes and thus Campbell's discussion of myths.

Stories reflect a ground-map of human nature and behavior, following a consistent collection of values and rules. Overcoming the Monster is prototypical, paralleling a development from immaturity to maturity, from incompleteness to wholeness and self-realization, something that can be described almost arithmetically, with the core family, light and dark hero and heroine, a trickster.

Some of the elements that Booker suggests having led to the departure from the basic plots in the last two hundred years have to do with the egotism of the hero or anti-hero; the story-teller's direct involvement in the story; and the inclusion of sex and violent scenes as such — sensational, but dead ends — in the stories.

Metaphors (which we will meet when discussing on how to communicate in a later chapter, but we will see more of it later in this chapter also) share with stories some qualities such that lessons or messages need not be spelled out in detail or very explicitly. They are easy to remember and it is acceptable if they have diffuse meanings. Metaphors may also be used to question or disrupt received wisdom but with less threatening impact; stories as well as metaphors can have many layers and be sophisticated — stories more, since they are less condensed. Stories help us interpret and understand actions and motives, thus we may learn, bond, remember, and share and embed emotions, persuade, inform, and inspire. Stories have power, embody norms, values, role models, culture, embody trust elements and tacit or silent knowledge.

Powerful stories may help reshape understandings and all-to-set views. But as stories are ambiguous, they can be interpreted in several ways — something useful in the beginning of a design project. Different interpretations help to test what framework applies, how the context should be understood. Stories help to raise problems and incompletely understood issues in an oblique way. They also help gain access to tacit knowledge, inaccessible to precise verbal phrasing. Discussing and developing stories help team members develop a common language. They also serve to establish team members on an equal footing.

Parallel worlds

What about analogy? Bookmarks constitute a concept transplanted from the physical world to that of Internet browsing; how might it transfer to a car, a hand-tool, or a wardrobe? Analogy is another way to achieve what stories do, or part of it, and metaphor does it, in a less formally constrained way than analogy.

One designer confesses to being taken aback at first when he heard a client suggest that "designers orchestrate the obvious".[12] The obvious? Obvious only after the fact! 'Organizational capital' is the catchphrase for capturing the fact that traditional investments alone do not suffice to explain differences in outcome for what overtly seems to be the same action — in fact, outcomes may be starkly different, tacit knowledge often one factor in the equation.[13] Thus, the designer may be called upon to take a systemic view, to orchestrate a system where 'soft' aspects may weigh heavily. In *Design-Inspired Innovation*, we borrowed an example from Marco Iansiti[14] who pointed out that during the downhill World Cup season when Franz Klammer of Austria won every single event, Klammer had the best intermediate time for none of the parts of any one race: he mastered the totality, never sub-optimizing.

This would be yet another broadening of the scope to encompass corporate activities, but it applies also to the product, service, or experience design. If an orchestra is out of sync, individual contributions, however brilliant, result in disharmony and disaster; conversely, in a harmonious orchestra individual members inspire and relate positively to each other, generating positive feedback, increasing marginal returns in economic lingo. So searching for that key component or expression that guarantees success may be a blind alley: perhaps rules and delimitations must be rewritten, like when IKEA decided that customers do the final assembly.

The parallel world may be very nearby: what about designing not for the product per se, but for the shadow it will cast? Inspiration may be visual, looking at nature or the artificial world — or looking for absence. Ponder a skeleton: nature 'drills' holes; it may be conceptual, when we take cues from understanding processes; and it may be computational, possibly through simulation. Simulation or not, in *design dance* a designer's body movements are translated into a design by a computer; the designer's movements may represent her interaction with a product, and the computer can deduce the product's form.

What about designing for animals? Think about it: design embodies culture and cultural assumptions, often relying on visual or language instructions for use. Domestic animals, and sometimes wild, may, in a sense, use products but without any access to these designed meanings — certainly a design challenge. To design also for animals may have a value in itself since even if we overlook it, animals share our space. In addition, it may inspire creativity. So, it is an extra challenge to see animals as end users, and to empathize with them — a pedestrian crossing, say, for hedgehogs, badgers, frogs, moose? It may be very difficult to understand enough to design for people from other cultures so what with animals not being able to share our culture at all?

With the advent of cell phones, countries with less developed or non-existing telephone systems suddenly could leapfrog development. No need for costly landlines; antennas and handsets serving perfectly as substitutes for traditional types of telephony systems. Cell phones in developing countries demonstrate a budding nanomarket for telephone calls: the village phone, hosted by some entrepreneur, often a woman, is used by anyone, who pays per call. There is a growing movement — including a corporate powerhouse such as the General Electric company and a design firm such as frog — acting upon the assumption that some novelty, some pioneering design, will emerge out of developing countries because conditions — constraints — are rather different there from what they are in the developed world.[15] Again, it is a change in perspective, in context, in demands, in boundaries. Frog found it challenging and inspiring to approach a market where they were positively ignorant — easier to pose those so-called 'stupid' questions.

Lessons from benchmarking could be applied. A service station at the Le Mans 24-hour car race must always remain on the alert, ready to cope with unforeseen emergencies. How does that compare to an intensive care unit?

One way to spur creativity is to change psychological distance, to create distance to what is occurring now, here, and to 'us,' thus making the issue more abstract. "Psychological distance" may be established by changing the ways of thinking about a problem, e.g., taking another person's perspective, rephrasing the question, or taking the long view, posing the problem in a future context (typically, in several studies, a life now versus 20 years into the future), alternatively, going *back* in time. A glass of wine may be a pleasant enough sensation, moving away perhaps evokes the picture of participating in the wine harvest, the European wine mountain, sorry, lake, bringing up troubling European Union agriculture policies. Or take results from Lile Jia's group at Indiana University.[16] On a concrete task, where students were asked to come up with ideas about a problem related to other students, one group were tasked with students within an organization on another continent, another group with students at a university nearby. Participants in the first group, thinking "distantly," generated more ideas than the "nearby" group.

Interestingly, Jasmine Cox[17] has designed a 'displacement engine' to provoke the establishing of distance in location. The user informs the 'engine' about her current GPS coordinates and it then guides, disguides as it were, the user to a dislocation, still offering a way back to where the dislocated walk began. In Cox's words, the engine breaks resoundingly with our striving for, well, comfort, or, to quote: "a beautifully uncomfortable object" and "a Displacement Engine is not your friend."

Eliminate distance: run a contest open for all, everywhere. Establish an open innovation gateway over the Internet. Court social media.

What about creating the future in a large multiplayer computer game? This is what the Institute for the Future did in 2008 with Superstruct,[18] an on-line forecasting game, lasting for six weeks. Together, thousands of players created new ideas and solutions to help postpone or avoid the 'end of the world' in 2042. As with good forecasting, several futuristic game solutions resulted in downright practical actions swiftly implemented. Another example is the game 'Building Futures' developed by Philips Design for a Dutch organization assisting companies to collaborate and shape future-oriented solutions. Here, teams design possible futures interactively and competitively, with ideas as stepping-stones.

Equally intriguing, as the future or geographical distance, perceived likelihood is a 'distance' factor too. If participants in a study think of a problem as highly unlikely ever to happen to them in reality, they are better at solving problems associated with it than if it is felt to be around the corner — we may say that they feel more detached if likelihood is low. An obvious prescription seems to be to travel to far away places, but it might actually be sufficient to *think* about such a journey or just the distant place. It enhances the argument that we should communicate with people who are different in as many dimensions as possible, hailing from different language backgrounds, etc. What about strange music from other cultures, exotic types of drama like the various Japanese theater styles foreign to a European, or exotic food and drink?

Design as animals are designed is something else than designing for them: it is about imitating biology, biomimicry. There is a vast amount of material available[19] to illustrate the power and potential of this source for design and, in fact, engineering advances, so, here, only a few examples will be given. The classic example is Velcro, invented or discovered in nature in the early 1940s by a Swiss engineer, George de Mestral. He took the burrs from thistles, found to stick to his dog's hair, and looked at them under a microscope: they had small hooks. A successful idea was born!

"Gecko tape" imitates the adhesive behind the lizard's trick to climb vertically, even upside down. An architecture bureau mimicked termite mounds for an office building. These mounds are self-cooling, and heating/cooling energy consumption was more than 90 percent reduced. The Shinkansen bullet train in Japan had a severe drawback in that it caused a thundering sound when leaving a tunnel, so the chief engineer, an avid bird-watcher, decided to have the train nose imitate that of the beak of the kingfisher. Result: much less noise, 20 percent less electricity consumption at a 10 percent higher speed. Figure 6.1 below shows a Mercedes concept car, imitating the boxfish, featuring an extremely low drag factor: fuel consumption is reduced by 20 percent,

Figure 6.1.

nitrogen oxide by 80. The question asked in both of these transportation cases, the Shinkansen and the car, was whether nature had solved the same or a similar problem, but the question may also be the opposite: what solutions have nature abhorred?

What are the design opportunities evoked by being inspired from nature's fractal structures or, say, meringue? Ticket dispensers at Arlanda airport in Sweden say 'plopp' when delivering a ticket, such as when a soap bubble bristles (no animal here, but nature). And here is a chair mimicking a bubble (Fig. 6.2):

Figure 6.2.

There is the Halcyon concept car purportedly designed to let the driver experience nature, therefore made to resemble an owl. The car body material has a composition similar to that of an owl bone — a rather odd form of biomimicry, no? At MIT, researchers found the human ear and its cochlea mimicking "a super radio with 3,500 parallel channels" so they designed a chip based on that concept, gaining ultra wide-band reception at one percent of the power previously needed.

These (Figs. 6.3–6.4) are cacti-inspired designs in...

Figure 6.3.   ... architecture ...

Figure 6.4.   ... and chairs ...

Cacti are "designed" for surviving in very dry environments. For a needy design, try the question where nature poses extreme demands, and what solutions have evolved.[20] Lotus leafs are resistant to wetting and when the water runs off, contaminants follow. The secret is small bumps on the leaf surfaces, promising to turn boat and car waxing or windshield wiping obsolete. The publicly most well-known application of biomimicry, together with Velcro, is probably the sharkskin swimming-suits, guaranteeing new sets of records.[†] Refractive paints are based upon extremely thin films, imitating the color effects from a butterfly's wings. No pigments are involved, and the color is dependent upon the exact film thinness. For an ice ax, designer Franco Lodato found nature's best hammer in the woodpecker (25 hits per second), which was translated into a shaft centered under the ax and the spine slightly curved (the bird rests itself on its tail), the pick pitched forward as the beak of the bird. The resulting swing is more balanced, and the blow more efficient.

When working on a theater play, the actors, guided by the director, experiment, try, and try again, rehearse, iterate, and then, in contrast to movie-making, confront a live public whose reactions are monitored. The similarities with design are obvious: subtle signals, interaction and interplay between roles and actors, stage and backstage, interpretation, light effects, music, and scenery, allusions (or not) to current events, to literature...

Why not create a foreign world? Designers from several design firms have learnt a lot from having to design for the Space Station: a cramped space within, a vast one outside, weightlessness, the necessity to keep track of minute tools, components, and details... What, then, of designing for a colony on Mars, in Asimov's Trantor or Robot world, in Farmer's Riverworld? What of designing cars as if they were cell phones or cell phones as if they were farming machines? Airplanes as if they were

---

[†]Beside the point, but the rules were decided to change from 2010.

pharmaceutical products? What if the product or service was a fruit, a sport, a specific author, an animal? Or a different trajectory than the one history took — IBM being the company saying yes to Chester Carlson's suggestion about xerography; Xerox using the PARC breakthroughs that became so plentiful and successful but chiefly outside Xerox, such as, among others, the graphical user interface of the Apple Mac and Ethernet? What if Bell Labs had not had such a foresightful manager in Mervin Kelly who brought together the three complementary scientists Brattain, Bardeen, and Shockley to beget the transistor, "ahead of its time"? What if said Shockley had not been so impossible to work with, something that caused eight gifted young engineers in his lab to leave to found Fairchild Semiconductor, a spin-off pattern they and their colleagues then repeated with Silicon Valley as one result? Such speculation may seem moot but could, with appropriate alternative trajectories chosen, generate discussions as to mechanisms associated with current design tasks and the context dependency of the resulting design.

Diversity

"Oedipus Rex" is an opera composed by the Russian émigré Igor Stravinsky who yearned to have a classic theme, and so opted for a two and a half millennia old Greek drama of Sophocles, asking the homosexual French poet Jean Cocteau to write the libretto and then, to underline the work's classic nature and not distract from the music, have it translated into Latin by a cleric. The American Julie Taymor (with four years in Indonesia and half a year into Japanese theater) has directed a prize-winning version with the English tenor Philip Langridge in the title role, the black American soprano Jessy Norman as Jocasta (she was intrigued to be directed by a girl, someone who was doing her first opera), the Welsh bass-baritone Bryn Terfel as Creon, and the Japanese Kayoko Shiraishi as the Narrator, speaking in Japanese; scene cast by Kazakhstan-born, Moscow-educated George Tsypin; conductor Sheiji Ozawa (who spent his first nine years in Manchuria), and musicians, dancers are Japanese — what a triumph for diversity generating stunning unity![21]

There are a number of possibilities for describing individual ways of approaching problems and of seeing the world. Psychologist Howard Gardner[22] has proposed that seven different types of intelligence exist (later, he added several more items to the original list). Ludwik Fleck[23] was the first to describe how thought styles define thought collectives and their languages. It has become rather popular to try to describe and to inventory variations between people's competencies, and in some firms there are 'tags' such as pins of different colors to tell what competency or thinking style 'tribe' a person belongs to. Then, there is the disputed Myers-Briggs Jung-based system, and the 'Belbin roles' evolving within a group context. People hail from different cultural backgrounds, and in the *Introduction*, we quoted research demonstrating that "treating chairs as masculine and beds as feminine in the grammar [turns out to] make Russian speakers think of chairs more like men and beds as more like women." Views of gender qualities are transferred to innate qualities of objects with associations also carried over to a gender neutral language. Likewise for depictions in art.

So: since diversity has proven to beget creativity, there are many ways to induce it. When IDEO spokespeople talk of T-shaped persons, they indicate a preference for those combining special competency with a broader knowledge. These might be people combining several unusual knowledge bases out of, e.g., science, technology, and some of anthropology (e.g., social anthropology), ethnography, economics, and also political science. Most probably there would be artists like a sculptor or a filmmaker present too, and someone who dabbled into the theater earlier in life...

The effects of words are many and surprisingly consequential, something to consider both when words are used in connection with designed products or services and when trying to avoid, perhaps, unconscious mental blocks to creativity. Langer[24] reports how psychologists John Bargh *et al.* set subjects to solve anagrams.[25] The control group had neutral anagrams, the other anagrams of words associated with old age (e.g., felorguft from forgetful). After the anagram session, the subjects' short walk to the elevator when leaving was measured. The group primed to old age walked significantly more slowly. Another study: people were asked to sort photos, one group putting them in the categories 'old' and 'young'. This primed again for slow walking. Asking for a regroup of the photos into categories unrelated to age eliminated the priming effect. Priming indirectly for health instead of age with, say, a language proficiency test designed to activate thoughts about a healthy lifestyle (or the opposite) affects whether the subjects would choose to ride the lift or walk the stairs.

To quote more examples from recent psychology research: power is associated with the direction 'up' — we really look physically up when 'friends in high places' are evoked. Holding a warm or cold cup of coffee influences the feeling about the personality of an individual a test subject meets somewhat later. It works in the other direction too: recalling a bad patch in one's life makes the room temperature feel lower. Touching objects covered with sandpaper makes people doubt that a social situation will be resolved smoothly (inducing self re-enforcement?). Moving pieces from a lower to a higher place makes people feel happier than when doing the opposite. Cleanliness and dirt associate with high and low morality, respectively; a literal washing of hands has the corresponding mental effect. Hardness equates with difficulty, so ideation should preferably be performed from soft chairs!

Therefore, metaphors may help designers get to grips with the new and vague or diffuse problem situations; they may function as stepping stones from the unknown into mentally safer ground. Similarly, Synectics relies upon analogy. Both metaphor and analogy are instrumental in elucidating the unknown and induce creative associations. Thus, if metaphors, expressed in language or physically, are powerful, why not try provoking creativity with mixed or contradictory metaphors or false analogies?

Wine tasting and efforts to describe wine tastes illustrate the limitations of language as well as the many ways out of this conundrum that have been tried. One element is the fact that there is no agreed-upon mechanism that explains the human olfactory system, which is key to wine experience. Here, metaphors, nay, illustrations truly abound. Fresh grass or black currents: these are pronouncements to convince, not to be monitored with a wine taste meter. Statistical analysis has mapped what words associate with high and low wine prices, but what direction causality? Revealingly, descriptions of expensive wines contain the word 'vintage' six times more frequently than those of cheap wines, which, instead, often are associated with the word 'harvest' on the label; 'steak' versus 'chicken' are meals suggested for the two categories. There is also the proposition that wine tastes find analogies in the world of music, an alternative to words but, in reality, not strikingly successful either.

The male and female bodies, and their reactions under certain circumstances, are different, which has had troubling consequences in the medical field. Ellen J Langer quotes more differences between women and men: women are less likely to die in the week before their birthday but are more likely to die in the week after.[26] But men are more likely to die the week before, with no deviation from the normal the week after. The importance on design is, in the words of David Jenkins (quoted by Langer), that men and women 'package reality differently.' Other facts with repercussions for design are qualities related to bones and muscles that are different between the sexes, not just size or proportions. More than men, and taken as a group, women look for benefits rather than features, for the whole experience of a product's function rather than just the product in itself. 'Taken as a group...' — but do not forget that there are vast differences between tastes as well as age groups and cultures.

A completely different sort of diversity is the one offered by trying to rely upon different materials. Artists see it as a challenge to make bronze look like wood, wood to be taken for steel, and so on. Coca-Cola in their new drink dispenser, described in Chapter 1, adopted techniques for micro-doses both from inkjet printing and medicine. Unconventional materials are associated with other costs and benefits than conventional ones, possibly allowing for new features and qualities, very possibly with additional costs and complications, such as new manufacturing or service demands. Design firms and design-conscious corporations often have some kind of 'materials library' or components showroom where enticing, inspiring, and tickling odd materials, tools, components, and tricks may be experienced. There are also companies set at furnishing such elements, such as Material Connexion[27] and Inventables[28] (these are just examples). Here two examples out of the repertory of the latter are:

- Temperature-dependent bar code, made from thermochromic ink, makes heat or cold to cause bars to appear and change so a scanner can detect the 'temperature history.' If, say, bar coded milk has been handled improperly, this would be discovered at the checkout counter. Currently, such bar codes are used for beverages and to guarantee safety for perishable products; it was first developed for a winery.

- Squishy metal parts look just like metal though they consist of silicone. The first application was a metal effect at the center of the steering wheel of a car: a visually metallic emblem but safe if the airbag has to be activated. Metal looks for running shoes have been suggested, as have (false) metal-ware for restaurants servicing children. Keypads for cell phones and goggles are other such applications.

- Britain's Design Council has provided some more examples: exotic two-dimensional arrays of particles held together just by light; zeolites employed as catalysts for cleaning car exhausts that promise to provide medicine with blood clot and skin infection prevention. And what about tailor-made skin and bones, for burn victims or disfigured patients? Or human cells suspended in a nutrient-rich liquid before being fed into an ink jet printer seeding three-dimensional structures being built up, resulting in tissue scaffolds?

This chapter dealt in creativity, a concept so popular that more than 150 definitions have been suggested. Partly, this is because it applies in so many fields, from poetry and sculpture to science and technology. In technology, we often contrast incremental improvement and *kaizen* with radical departures. But might there not be several flavors of creativity? May such contrasting simplifications lead us astray; may they risk misleading us?

Creativity researcher David Henry Feldman urges us to remember what he refers to as Middle C, in an allusion to the musical note of the same name. In other words, we tend to talk about Great Creativity, big ideas and radical breakthroughs, or (if we refer to them at all) we talk about the opposite — small creative incremental jumps — as more of an aside. This is despite the fact that, when taken together, many small leaps can make a real difference.

The point of that Middle C is to remind us that we are dealing with a continuum — a continuous, ongoing scale — where what is radical to a greater or lesser degree is in the eye of the beholder, determined by the observer's outlook or vantage point. If a researcher or practitioner focuses on either of the end-points of the scale, there is the risk of being led astray. In reality, how often does an idea actually reach the absolute extreme — high or low?

To refer to innovation instead — radical as opposed to incremental — is, of course, no different at all if we stick to creativity applied to technology or business in a broad sense. Quite reasonably, business idea or innovation competitions encourage what seems like big leaps, while work relating to improvements and sometimes suggestions for improvements — again, *kaizen* — is the preserve of smaller-scale efforts in terms of radicality: middle *i* innovation.

After all, no-one is on the look-out for small advances, no-one competes to achieve modest innovation, and no-one receives recognition for a level of inventiveness that, despite not being low, is still only moderate: mesovation (micro, macro, *meso*). Why so? Do we not, then, run the risk that the best becomes the enemy of the good? Why not also reward mid-level creativity?

The same goes for those of us who are creativity and innovation researchers, and for those who study our/their works. By the very nature of things, it is the radical, attention-grabbing breakthroughs that end up in the spotlight. But what do we thus miss out? What do we lose in terms of a more general understanding that could be applicable at both high and low levels, both at the highly innovative and incremental ends?

As we have seen early on in this book, different languages "see" the world differently, affecting how we see the world also. The Swedish language actually has a word for that which is just right, in the middle — *lagom* — a word lacking a satisfactory translation into many other languages. It would hardly be appropriate to call for an ombudsman or a champion for all that is *lagom*, to inventory the degree of lagom-ness, but wherever an adequate solution is lacking, a niche arises for unique competitiveness. Within the market for goods and services, Middle C and mesovation could be a factor for gaining a competitive edge, while not something that creates its own specific industry. When it comes to research and education, however, the situation is quite different. If a field has not been studied, there are plenty of opportunities for making discoveries that bring their own reward — cited articles, titles, positions, and grants that belong to the realm of scientific endeavor. Think, then, about design — which introduces another dimension to that continuous spectrum of creativity.

Partly the idea behind this line of reasoning came from studying those two books by Umberto Eco: *On Beauty* and *On Ugliness*. However, there are *no works* about *the middle ground* — all that which is indifferent, lukewarm, moderately interesting, or even uninteresting and tiring. But note that Middle C and the mid-level of innovation — mesovation — need not be uninteresting or indifferent; they simply belong somewhere closer to the middle ground when viewed on a continuous scale. Instead of lukewarm or indifferent, try, if not *lagom* then *appropriate*. So do not forget to celebrate moderation sometimes — that which is *appropriate*!

Endnotes

[1]A nice overview is found here: http://www.slideshare.net/crafitticonsulting/lean-inventive-systems-thinking-work-book.

[2]Olsson, Bengt, 2008. Beskrivningsspråk i och för kreativ praxis: Idéutveckling under gruppsession. Doctoral dissertation (in Swedish), Mälardalen University, Eskilstuna.

[3]A Dialog seminar manual may be downloaded from http://www.dialoger.se/sida.asp?rubrik=44.

[4]In Olsson's thesis work above, one of the texts relied upon in the series of Dialog seminars was an excerpt from Hofstadter's book on translating/interpreting. Hofstadter, Douglas, 1997. *Le ton beau de Marot*. Basic Books, NY, NY.

[5]Henwood, Esther, 2009. *Design & Littérature*. Norma éditions, Paris.

[6]de Bono, Edward, 1985. *Six Thinking Hats*. Little, Brown and Company, NY, NY.

[7]http://www.design.philips.com/philips/sites/philipsdesign/about/design/designnews/newvaluebydesign/july2009/playful_innovation.page.

[8]Martin, Roger, 2009. *Design Thinking: The Next Competitive Advantage*. Harvard Business School Press, Boston, MA.

[9]http://www.savethewords.org/.

[10]Manguel, Alberto, 2008. *The City of Words*. Continuum, London.

[11]Booker, Christopher, 2004. *The Seven Basic Plots*. Continuum, London.

[12]http://www.fastcompany.com/blog/steve-mccallion/beyond-widget/creating-consumer-experience-innovation-building-value-metaphors.

[13]Brynjolfsson, Erik & Saunders, Adam, 2009. *Wired For Innovation*. The MIT Press, Cambridge, MA.

[14]Iansiti, Marco, 1998. *Technology Integration*. Harvard Business School Press, Boston, MA.

[15]http://www.businessweek.com/magazine/content/09_12/b4124038287365.htm.

[16]http://www.scientificamerican.com/article.cfm?id=an-easy-way-to-increase-c.

[17]http://www.jasminecox.co.uk.

[18]http://www.superstructgame.org/.

[19]The Biomimicry Institute is a good source: http://www.biomimicryinstitute.org/, http://www.asknature.org/, and Janine Benyus herself http://www.ted.com/talks/janine_benyus_biomimicry_in_action.html; http://hbr.harvardbusiness.org/web/2009/hbr-list/business-of-biomimicry.

[20]More at, e.g., http://www.technologyreview.com/biomedicine/23933/.

[21]It is available on DVD.

[22]Gardner, Howard, 1983. *Frames of Mind*: The Theory of Multiple Intelligences. Basic Books, NY, NY.

[23]Fleck, Ludwik, 1979. *Genesis and Development of a Scientific Fact*. Chicago University Press, Chicago, IL (translation from original German, 1935: Enstehung und Entwicklung einer wissenschaftlichen Tatsache).

[24]Langer, Ellen J, 2009. *Counterclockwise*. Ballantine Books, NY, NY.

[25]Bargh has made a whole series of experiments in the same vein.

[26]Langer, Ellen J: Op cit.

[27]http://www.materialconnexion.com/.

[28]https://www.inventables.com/.

Chapter 7

CHANGING AND BROADENING THE SCOPE

Frequently, an important designer contribution is to challenge the design brief given, starting afresh instead of just designing as per request. One way of offering a challenge — or, as we shall see, in effect, several ways — is by broadening the scope and thus changing the view, putting the task in a new context, reframing it, and possibly to change the entire point of departure. Broadening the scope — or moving laterally, to something unexpected?

To rephrase it, rather than look for an idea or ideas, the fundamental resource might be directed at getting to the problem, to the root cause of a worry or a need. So, why this particular request for a design, what is the ultimate reason, the objective? Why look for a drill when what you need is a hole, nay, a bookshelf affixed to a wall? Or might the bookshelf be substituted, or even the books made redundant? Or, perhaps, books plus shelf serve as a symbol — that may be substituted?

Infinite regress is not very practical but what might be productive is posing such questions to open up a broader set of alternatives and ways of action. It may also be highly fruitful to ask the question: what are the *real* constraints, what do the laws of nature really prohibit and thus what do they allow, as compared to restrictions and traditions that are tied to the human mindset, and, therefore, very possibly the results of historical circumstance, opportunism, immediate pressure at a certain instant, haste, laziness...

Such conditioning is often the root of rules of thumb, and in addition there are rules for solving certain problems that reflect an ancient state of the art when certain laws of nature had not been discovered or mathematical tools invented or when, for example, computer power put strict limitations to how problems might be treated. A lateral way of changing a boundary is simply to skip a certain constraint to better understand what it implies, or to do the opposite, introduce some non-existent restriction, to learn how to work around it.

Whatever product or service, there is bound to be several competitors on the market. But do those constitute the *real* competition, from the putative user's point of view? For radical novelty, the problem is often to convince the customer that the novelty is at all viable — or to have the customer understand its functionality. In such a case, history shows that it might be an advantage if there are several competitors establishing product category credibility together. So — what *is* the real competition? How should the broadened scope be envisaged? Is that competition for time, for money, and if so, for what 'compartment' of the customer's wallet? Or are there learning barriers? A new kitchen may compete with a new sailing boat for a household's budget — a pair of running shoes with the latest weight-loss fast food.

Again, broadening the scope or sneaking up on the problem from an entirely new and unconventional direction may result in profoundly different — novel — solutions. Of course, designers have no exclusive talents for taking a broader view so non-designers should feel encouraged to rely on much the same tools for thought. Designers, however, already by the fact that they often approach the problem-solution duality from an aesthetic vantage point — among others — are bound to offer one or several alternative approaches. Furthermore, to reiterate, they often and quite naturally broaden the question by phrasing it as "what exactly are we going to design, within what framework, and how is that framework designed; do we need to redesign, perhaps, the very framework because the constraints that are imbued with it raise obstacles to good solutions"? May we need to invent or design an entirely new concept, venturing into *concept design*?[1] Moreover, since designers tend to move between industries, disciplines, and technologies, they are thus bound both to embody very deep and detailed competencies and, on the flip side, to master a much vaster gamut of disciplines, techniques, and technologies.

This idea may be generalized to encompass just anybody from the outside, coming from an odd direction. Children would not be identified as designers and certainly not as professionals either. There are, however, numerous examples of children taking original tacks which resulted in profound insights and delightful ideas, even if not offering straightforward solutions.

A larger scope, designers would tell, can result if clients allow them to sketch scenarios and design for environments beyond the time horizon of the current project, thus provoking alternatives and sensitivity to longer-term developments that may have high value. The end result may well be unusually long-lived solutions. Therefore, it is best not to short-change the work of designers. Thinking about designing the future forces valid questions about the viability of today's solutions, and there are already decisions or, at least, signals about new regulations and demands that are bound to arise. While designers as well as corporate planners have a hard time generating long-term forecasts for the world of information technology, other societal and economic systems, such as those involving energy generation, are decidedly slow-moving and sus-ceptible to forecasting for ample periods of time. So what about energy consumption patterns ten years from now, and what about environmental concerns? What about demographics and consumption trends, a graying population in the industrialized world?

Or what about changing the system boundary? This is what happens when companies that are not look-alikes merge. What about a merger of ideas and viewpoints, engaging with partnering organizational units without any corporate merging? Establishing partnerships with other companies offering complementary products so as to create a new whole product together obviously makes for changing the boundaries and the entire system. In-house or consulting designers who have worked with other clients from other industries may contribute a kind of virtual merging of perspectives.

So far, we have dealt with a change in scope of two varieties, one relying upon *persons* with different faculties and perspectives (e.g., children, as opposed to adults), the other on a broadening of the *system* considered.

One example of how re-defining a systems boundary may be beneficial is the garbage treatment station discussed in *Design-Inspired Innovation*. Among competitive bidders, two architect firms suggested look-alike buildings, garbage treatment stations 'as usual.' The third bidder, employing industrial designers, first undertook a functional analysis and reconsidered the entire waste management system, including, for example, the trucks that were to deliver the garbage in a sensible way, and the surrounding environment that should be saved from pests like birds and rats scavenging on the garbage, littering and disturbing the environment. What was the basic purpose? What were the different dimensions, aspects, requirements? How were the waste containers to be serviced? What were the environmental conditions and demands? How might the system look and behave after three or eight years?

The research phase included travelling with the truck drivers as they picked up waste and then unloaded it at existing stations. The end result was a suggestion for a modular arrangement with movable containers placed within concrete structures. Instead of the traditional large trucks, small lorries could be used to access the garbage station more easily. The aim was to produce no waste at all except what was contained in the structures, and also to minimize the risk for accidents. With their concern for the environment, the designers had, in fact, broadened the concept of the end users beyond waste management, to include people who see the site and live close by it.

A broader scope or changed perspective has several effects. One fairly obvious result is spurring creativity through the fresh angle of attack, through the questioning of 'business as usual.' On the next level, when the broader scope is brought to bear on a project or on several projects, users and customers are likely to receive a refreshingly novel result, offering something extra as compared to that of 'business as usual.' And for the provider of the designed object, service, experience, network (linking, perhaps, to 'the network of things'?[2]), etc., the broader scope makes for elements of unique sales propositions, for establishing a distinct profile relative to the competition.

In other words, asking the question 'what is the *experience* we would like to induce' means going out in the field to study how and what experience(s) play out now. What and where are the exact touch-points of that experience, the 'moments of truth' in a service? What seemingly tangential factors to the offer — product, service, or combination — are important, perhaps, *the* most important? What *emotions* do we want to evoke, or, is it rather the case that some emotions should never be evoked? Would we like to, and succeed in, getting the user feeling positively *transformed*? Design as *sense-making* — designed objects as *agents* — for what? — or *events*? Designing emotions, designing *for* emotions...

Designing for *surprise*? Yet another way of changing the viewpoint would be to opt to surprise the user. This might be done in two ways: misleading the user about what to expect so the experience is a surprise since the expectation is unfulfilled, or by creating uncertainty so that the user does not really know what to expect. In the first case, appearances can be outright misleading or conceal the real nature of what to expect, or characteristics may simply be hidden. In the latter case, feature combinations may be sufficiently original, shape or materials unknown and exotic so surprise is, in a sense, expected — but not its exact nature. Expectations, quite naturally, come out of past experiences from products or services similar to the one considered. A misleading surprise has a higher 'surprise value' but is also more likely to be perceived as negative precisely because the user feels betrayed: low perceived uncertainty has been transformed into a feeling of being gulled. If the surprise factor is tangible, uncertainty is perceived as large from the outset, and the genuine surprise is less — a surprise is expected, its exact nature an experience. The level of surprise has no correlation with how the user judges the beauty of the 'surprise' object.

A system has to abide by Ashby's Law of requisite variety, matching the complexity of its objective. Changing and broadening the scope brings about questions regarding meeting this requirement as well as what may happen to the context for what is being designed. What should be included, what not: what constitutes the boundaries and the limitations? Systems thinking is about just such issues, a system seen as having an external environment to which the system has an interface, and, internally, consisting of subsystems and components, with internal relations that all together make up the system and its behavior. Thus, a system on a lower level may function as a sub-system or a component to a system defined more broadly — the scope enlarged. Even changing the internal structure of a system changes the system as a whole, and to do so, it is sufficient to change those internal relations. New communications protocols or new software may, therefore, bring about major changes in hardware systems that are physically the very same. The context of an interactive system must not be seen as one single entity.

To reiterate, we emphasize that changing relationships within the system, changes in interfaces and in rules or software, bring about changes in the system as such, even if not in its physical aspects. In the real world, almost always, multiple users, not just one or two, interact with any system, and if in the process, as a result, they interact between themselves or influence the system behavior, the very nature of the context can be said to change. We may see the context as changing, or the system as having multiple contexts, each one linked to a particular ensemble of users and use of the system. Furthermore, we may think of physical space as different from information space, which, in turn, is different from media or interaction space.

Yet another design tack would be to venture into *designing contexts*, or, rather, context platforms. A special type of systems thinking is the one provided by Systems Dynamics. Here, the world is seen as consisting of *sources*, where something is generated, *sinks*, where it is consumed or destroyed — both outside the system as such —*flows*, for transportation within the system, and *stocks*, where flow content is stored. Time is, of course, an essential variable. As Senge illustrates — and as Forrester pioneered — in, for example, The Fifth Discipline,[3] the delays and mismatches between goods and information flows may be the cause of grave misjudgment, the point being that it is all systems dependent: the *architecture of the system* makes many seemingly obvious interventions irrelevant or produces counterintuitive results — what looks like wise decisions is just futile, and at times even counterproductive. 'Common sense' makes no sense without taking systems structure into account. So *systems design* is of paramount importance!

Designers often take dreams and visions as their points of departure (though they may not necessarily phrase them as dreams). The challenge raised is phrased something like: "Wouldn't it be just great if ...?," forgetting for a moment limitations of various kinds — including any original design brief. Thus, the scope is broadened to encompass the impossible, to turn the impossible possible, or to return to the possible by another route. We saw how the designers developing the KOR bottle put the question to Eastman Chemicals whether there were any alternatives to polycarbonate and received the answer: "Not quite yet." And somewhat later, the impossible had become possible.

Moving laterally, changing the scope but perhaps not broadening it, might entail changing the meaning of and associations linked to the design. Should home electronics look like coming out of a high-tech laboratory? Or like pieces of furniture? Instruments from a surgeon's clinic? Like toys or resembling computer games role figures? Characters out of commedia dell'arte, or figures taken from Tarot cards? We recall the Merry Spanish furniture in our Chapter 1.

The most broad-ranging scope in the time dimension would encompass the entire life of a product, 'from cradle to cradle' in the famous words of Walter Stahel, at best, mimicking nature in the sense that loops would be closed and everything recycled. Previous concerns for taking into account ergonomics and energy savings may be seen as forerunners. Several countries, including the Netherlands and Sweden, have carried out programs where industrial designers have been equipped with tools for performing product life cycle analysis for customers, followed by a redesign of products with the largest potential for success in achieving reduced or even zero environmental impact. As with other ways of shifting the view, new ideas, some of them little related to environmental issues, appear.

In several cases, *design for a closed life-cycle* turns out to be outright profitable in itself, in some requiring a change in, for example, materials thus demanding existing production machinery to be substituted, for financial reasons perhaps at the end of its economic life. In some instances, to be sure, changing to a design with more concern for sustainability is costly and may be motivated instead by its goodwill effects on customers and other stakeholders, e.g., employees. Yet another possibility is that something unprofitable will turn worthwhile later, when regulations get stricter. Often enough, a thorough analysis of the entire chain of events that brings the product from when it does not yet exist to when it has been entirely decomposed generates in itself powerful insights and inputs resulting in better design. Change that chain of events, maybe?

For the development of an environmentally friendly commercial cooling display, Electrolux leaned on Dutch DfE software. The disassembly time for the product was thus reduced by 40 percent, 96 percent of materials being recycled at the end of product life, and energy consumption reduced by about a tenth. The design included:

- Foam strips easily peeled off replacing silicon insulating strips; the new strips also looked more attractive
- Larger copper evaporators for improved energy efficiency
- Replacement of materials to ease recycling, e.g., polyester bins filled with polyurethane were substituted by a combination of recyclables
- Aluminum and copper content reduced
- Modularization easing disassembly and reutilization of functioning modules

The British Design Council has, for its compilation of design statistics, suggested the following distinctions that may form a useful backdrop for the subsequent discussion:

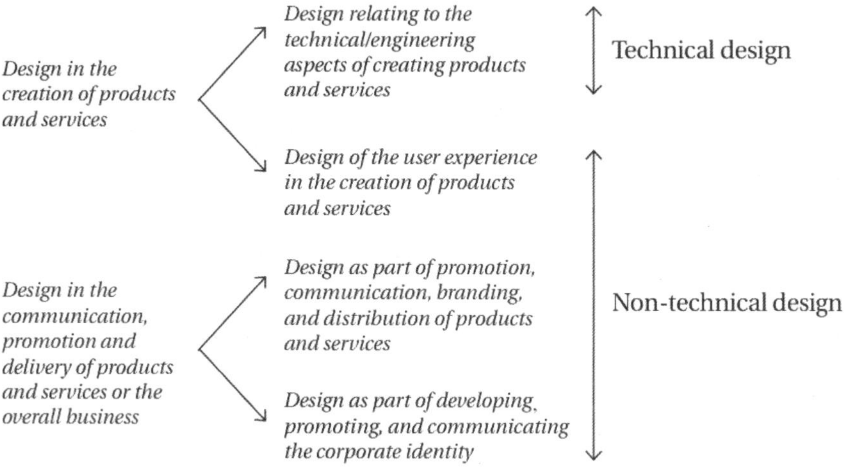

One particular type of changing the scope is when a company decides to outsource its entire development effort, its innovation activity, to a design firm. This is the case for companies having taken a hard look on what are their strengths and weaknesses, and who have decided to concentrate on these, *outsourcing* everything else. Of course, several types of such concentration exist, including when competitive design firms are invited to suggest designs for a company, concentrating, rather, on *judging designs* and their innovative effects than developing anything themselves. Much depends on what the strengths are that the company, a customer to the design effort, has decided to concentrate on. If it is a certain production machinery and manufacturing process, possibly very capital and/or competency demanding, then that provides a distinct focus for designing; if it is a company that excels at marketing and selling, then that fact establishes some guidelines for the design efforts. Obviously, market insights and end user reactions are keys — often unknown keys — and the relationship between the design firm and its customer company has to be fine-tuned.

For a company to outsource something often described as a core activity is a far-reaching strategic decision with long-term repercussions. It would need much reflection and a well thought-through design relationship in general, and formal agreements in particular. One of the advantages with such an arrangement would be that the design firm might concentrate on streamlining the development process, thus furthering, possibly accelerating, innovation. Another potential advantage is the crosspollination between projects for different non-competing design firm customers, from different industries. There may also be certain resources that make

the design firm a better development lab than the customer firm —
hardware tools, software, competencies, all of which can be profitable
only if employed sufficiently and where the design firm may achieve
economies of scale unattainable to the particular customer company.

Innovation process design involves design firms going beyond design to
function also as management consultants, implementing for a client what
designers have learnt about applied creativity and innovation work. The
design consultancy would not pretend to offer traditional management
or organizational design services, but stress its ability to demonstrate in
action how innovation projects should play out. The fact that emphasis
is not on traditional consulting, but leaves traditional management and
organizational tweaking to the client firm's management might make the
latter more susceptible to what it perceives as complementary. A design
firm may, again, act as a transfer agent between client companies from
different industries. Acting and demonstrating may also take the form
of giving courses oriented at design-based innovation, providing broad
and deep training programs, and providing written instructions on the
maintenance of design practices, creative work, and innovation efforts.

We are moving towards a third dimension of 'scope,' after the
individual seeing and perceiving qualities and the broadening of the
design concept: designing the entire company in some way. Outsourcing
was a step, redesigning organizations and modes of working are others,
and, as we shall see, there are several more.

In the case of the KOR bottle in our prologue, RKS designers kept
tabs on new materials development, and apart from being able to transfer

technology from one industry or field to another, designers often try to keep tabs of what promising or challenging novelties scientists, materials providers, and, for instance, software developers are up to. From time to time, designers may embark upon stretch projects without any client, not for profit but to learn about exotic technologies and new modes of thinking. Advice on materials choice is sometimes tangible and can turn out to be very important; an advice with experience and tacit knowledge attached to which the client would not have arrived without the designers' assistance. A drastic change, from, say, metal to polymers, entails new production processes, new logistics, rethinking services and maintenance, spare parts warehousing and delivery, quality measures, and possibly more. To generalize the point, design firms may be seen as engaged in *value chain design,* yet another broader scope that makes for questioning that 'business as usual' — evaluating and choosing between materials to be used in an innovative product, locating and evaluating suppliers for materials, machinery, and possibly sub-contractors for all or part of manufacturing. New incisive questions are bound to arise — what criteria should apply, what are core undertakings, and what can be outsourced, under what conditions?

Another view of the value chain than the one of sourcing, logistics, and production is also involving the corporate 'superstructure.' We may see the *business model* applied as an abstract structural factor, networks sustaining the operation as virtual structures open to design and re-design; moreover, any organization is, of course, dependent upon its economy; financial arrangements are open to design too, at least to some degree. Processes may be divided, traditionally, into core and supporting ones, again with the challenge of whether their balance may be shifted or something which can be outsourced. The product proffers core functions and performance but should also be seen as part of a product system, an ecology if we like, and in both of these dimensions design is possible. A third one consists of the services concomitant to the product — if it is not in itself a service. Finally, the product must be delivered, which requires a distribution channel or channels, and, associated with it, communication to convince buyers and inform and instruct users — and so, most often, the product is branded.

With combinations of freeware, shareware, freemium (basic service is free, with premium service at a cost; or the product for free — the manual costing a hefty sum), pay-what-you-like (yes, this has proven to work for products with high perceived quality and under certain conditions, when it may even prove more profitable than pricing; cf shareware and donation ware), and traditional models (subscriptions versus pay-per-issue) just for paying, it would seem as if the number of potential business models has proliferated from the time of the shaver plus blades. Perhaps; but whatever trend, another broader design aspect concerns designing, again, the *business model*. What we just saw was a number of payment models, including the zero payment option; we should, on the economy side, add revenue model, profit model, and possibly more. When it comes to the broader business model, opportunistic and long-lasting partnerships might come into the picture, as would, of course, customers and also customer relationships: episodic or lasting? What is the value proposition to the latter, and how may it be redesigned? An education system for the customer?

Considering a product's value chain, there is yet another important factor to take into account: the *value network*. This is closely related to the concept of *product ecology*.[4] An illustrative example is the introduction of the TiVo on the market, later emulated, a bit differently, by, among others, Canal+ with its Cube (see box). In the traditional value network, content and ads were integrated by TV networks, subject to some kind of regulation, in the US, for example, by the FCC. Consumers met the programs by way of TV sets produced by electronics companies. The TiVo and its counterparts inserted itself into this network, and brought a number of new actors and functions into it, like movie studios, providers of streaming video (Netflix, etc.), and auxiliary electronics gadgets (consumer electronics, computers) in addition to the TV set.

> Looking at the larger picture — TV experience: The TV chain Canal+ wanted to create a new dialog with its subscribers, and to that end had Le Cube designed. This is a HD capable set top box with a difference: the TV set, the remote, the dual-tuner satellite receiver, the PVR with a 320 Gbit hard disk, and the graphical interface are seen as aspects of the same experience. This integrated system brings information, personal preferences, and Ethernet communication to the subscriber even before the TV has been turned on. Of course, Le Cube features (minimalist) interaction by way of the Web; a media platform integrating television; broadband Internet; and information. The device even has a gyroscope, allowing it to sense whether it should rotate its content.

Apple, with its iPod, iTunes, and iPhone, is often quoted as having established — designed? —a *business ecosystem* by combining developer tools, media relationships, marketplace and e-wallet functions, one-click software distribution, platform governance, and a simple, compelling way of sharing revenue with developers and providers of software. Here, lots of synergies play out. iTunes is now the world's largest source of music, and the iPod is the base for a plethora of accessories, some produced by Apple, most of them by independent suppliers. We have seen how the iPod may interact with a product from another walk of life like the Nike+ shoe. The iPhone comes accompanied by its App Store where a gazillion — at the time of writing, more than 100,000 and counting — of applications may be downloaded. Some of these are free of charge and open source, while some are commercial.

When discussing designing the value chain, we are into the physical aspects of the product, and how it is made. Starting from the end user perspective, and particularly the user interface, we arrive at the functional performance or what the product does. Here, we also attach the sensorial aesthetics, the exterior, the impressions, and how it looks. One aspect of this is the user interface and, thus, how the product interacts with its users. But there is a more qualitative aspect to how the designed object feels: the emotions that it generates. How do we design for emotions — what emotions to evoke, what emotions to avoid? Moreover, there is the mental dimension, or perhaps several ones — the product might be imbued with cultural and behavioral meaning, and possibly with ideological and religious as well — consciously or unconsciously.

One way of broadening the scope is to engage in developing a corporate identity, in shaping a comprehensive *corporate image design*, including brand or brands and the concomitant logotypes, stationary, publications, exhibitions, web designs, possibly architecture and indoor design, and more. Taking a broader perspective is sometimes highly conducive to innovation: is it the brand that sells the products, or the products that establish a dependable, attractive brand? For industrial goods, they may serve as long-lasting posters for a brand and its maker, and the same holds for household goods that we fall in love with and make us eager to acquire more products with the same visual, audio, tactile, or whatever attractive qualities. The architecture of a building, with its adjacent parking places and access routes, cannot be easily changed in its basic qualities so this requires profound reflection, possibly to make it easier to undergo superficial change. Web designers re-discover truths about how to present graphics and pictures, but they also have cause to discover more about Internet surfers' behavior where click-throughs are important and attention is fleeting — discoveries that may spill over to other fields of user interaction and interface design. '*Corporate identity design*' can be a step along the way to strategy design (strategy design will be discussed presently).

A brand is sometimes identical with corporate identity, sometimes one of several branded offerings from the same firm. Widening the scope to *design a brand* could well imply that the brand embodies the product or vice versa: Swatch is a brand, and we identify it as a line of watches, individually different but with a strong family identity, and a strong brand. The classics among brands were, of course, the General Motors five-some, each corresponding to a perceived market niche and the concomitant designs and price levels (and dealer networks) attractive to those niche buyers. One way of thinking of brands and branding is to regard them as designing for persuasion, and advertising is an obvious element of that. Brands, however, can be said to be about fulfilling promises. Those promises should be kept when the customer encounters the brand, the design, which happens at touch-points. If we have talked about a value chain in production and logistics, it might make sense also to talk of a *delivery chain*, focusing on those 'touch-points' where the customer encounters the product — in various ways. And as with the value chain, that delivery chain may be susceptible to being designed: *designing for touch-points*.

Branding is important; respected brands command a premium in the market. Trend analysts PSFK have an expert panel producing an annual ranking of the world's 'good' brands, and an analysis of the 2009 list showed the following factors to be common traits of the top brands[5]:

Utility: Enhanced usefulness for the consumer, not only for the product or service, but the eco-system around it.

Experimentation: Constant innovation, continually pushing the boundaries of the offering, creating ancillary products.

Design: Premium aesthetics combined with consistent delivery, making the audience feel valued, and encouraged to adopt the product/service as part of their identity.

Community-listening: Creating a sense of community among customers, actively engaging them, and listening to what they have to say.

Change the model: Understand the customers' system of needs, and changing the business model to suit them.

Beyond the 30 second ad: None of the top 100 largest advertising spenders made it onto this list. (Might advertising even be counterproductive, or negative?)

Environment: Brands in the list's lower half lose points by fostering innovation over environmental responsibility: sustainable practices are a must.

In addition, the micro level of brands matter. Crayola found that 'colorful' names like Cornflower yellow or Kermit green worked much better than the bare color tags. Abstract connections seem to work better; the more abstract, the better.

Yet another entity to consider designing is a *portfolio.* A corporate portfolio would be the business units belonging to that corporation (think GM's classic five makes again). A business unit may contain several business, comprising another portfolio at the next level, and at the most concrete level, there would be one or several product portfolios. Much of the considerations pertaining to portfolio design would be financial, but not necessarily all. In fact, the customer should, as always, be considered, and a strong portfolio cannot stay strong without change, which is where design comes in. One might see *portfolio design* as a complementary view to other design concepts that apply to not only single products/services, but to whole sets of those. A portfolio attests to the fact that there is a demand, however transitional, and on that basis, the job now performed through the offerings might be done otherwise; there might be tangential or adjacent products or services to offer; and there might be gaps in the portfolio. Another category would be novelties that have no current counterparts in the portfolio or in any competitors, innovations that invent their own market, thus enlarging the portfolio. Portfolio design straddles the dimensions of broadening both product and corporate design; there are, of course, interdependencies in the three-dimensional scope universe sketched here.

In a certain sense, the most ambitious design scope on the corporate level would be the design of a *corporate strategy,* which would entail a number of profound change undertakings, including innovation efforts. Strategy designers would equate with strategy and change consultants. Issues such as corporate culture and other important historical impediments — geography, physical factors, reputation, and so forth — which define the client's "identity" must be mapped and understood. Probably that identity, including corporate mission, cannot be fundamentally redesigned but perhaps tweaked, and then concretely made into products. Once a (new) strategy has been defined, emphasis moves to design not just of products but also, for example, of packaging and even environmental qualities pertaining to the client, such as architecture. The designers must also work on product distribution, finding the most appropriate channels for a particular project/product. When engaged along this strategy avenue to innovation, designers have to be involved in positioning, in structuring the product or in new service launches, and in resource research. Product and service design are subsidiary to the designed strategy. Because of the span of this scope, a whole *network* of independent specialists have to be engaged, a team that can be said to be designed since they have to be able to work together, do it timely, and with appropriate expert competencies.

The ubiquitous Internet has spawned cloud computing which means that more and more of resources, from software applications to information such as documents and pictures, are stored and available over the Net rather than being stored locally. The role of the Internet as a market mediator is constantly expanding; it is an instrument for establishing and maintaining on-line communities and also extensions to non-on-line communities. It is of import to provider-customer relationships, to be sure, but the point of customer relationships brings us also to *their* internal relationships, that is, may we see them partaking in a community, related to our offering? We should not pretend to design such a *community* but, rather, to *design a platform/platforms* for it/ them, designed to be adaptable to the wants and actions of customers themselves, for communities spontaneously emerging. Or, maybe, there are communities that already exist for us to latch on to, or for our community platform to interact with? User communities are often enough the fountainheads of innovation, with their own inventions, solutions, and designs emerging out of their needs. They can be corralled to volunteer as beta-testers as in the software industry, and they might be sounding-boards for new products or tweaks on old ones. There would be no community without interactivity, so there must be a bulletin board for free discussion, including sharp reactions to the company's offerings.

It is a managerial task to see to it that the web site and e-mail-function are manned to provide prompt answers, but it is a design component to make that suggestion — underlined.

What about involving, engaging customers, *design users, in the actual process of design*? One level is that of pioneers or lead users engaged in a club, a tribe, a community, or a network? Another tack is co-creation on a broader scale, ideally involving all users or all that are interested. Yet another would be to leave the last moments of the design for the end user to create herself, to leave the designed object unfinished. What about designing so that the 'final' design is *not* the final one resulting but what you deliver is something that may be changed and shifted, as the user gets more accustomed to it, wants variation, tries to keep up with new developments or plain fashion trends? Some of this is happening when, for example, web sites allow users to choose between different skins, RSS feeds, and so on — at least to a certain extent. Some sites have actually started offering tools for the user to create her own skins, appearance, etc. SourceForge is an immense gateway to a myriad of open source freeware, designed by enthusiasts, and improved by other enthusiasts. Lego introduced a design system software for creating new Lego combinations and saw it tweaked by users. Instead of crying foul and start protecting immaterial property rights, they welcomed the tweaks and advertised their availability.

How may we design a system that ...

— ... caters to people's yearning to be generous, altruistic?
— ... attracts those caring about the environment?

- ... provides real beauty — and what is strikingly beautiful in the mind of your customer?
- ... thrives on/in the long tail?
- ... strikes the right 'freemium' balance?
- ... links up with the 'right' non-profits and NGOs?
- ... implies a new, compelling concept?
- ... links into the appropriate existing on-line communities, and creates (a) new one(s)?
- ... makes users themselves initiate and sustain a 'users' community'?
- ... strikes the right notes of presence at, for example, Flickr, Picasa, Slideshare, Youtube, Wikipedia, ChangeThis, etc.? Uses Twitter, LinkedIn, My Space, Facebook?
- ... has access to try-out-points with consumers, for experimenting and early learning of reactions to novelties?
- ... defines luxury or exclusivity without the associated wastefulness or exorbitant price?
- ... thrives on existing or newly invented touch-points, engrossing to the customer—user?
- ... gives you control over your company's and products' stories, creating timely prebuttals if ever something negative starts to spread?
- ... changes existing business models, rather than, say, selling cars, selling transportation miles or acceleration?
- ... takes the time horizon into account in new attractive ways?
- ... caters to the new demographics evolving?

— ... establishes harmony between features expectations and actual capabilities?

— ... in itself makes for a dynamic change in system boundaries?

— ... exploits what's next in location (GPS, Google Maps...), time (Twitter...), and more?

— ... makes for an ongoing conversation with the customer(s)? (including having them as co-designers)

— ... operates like a software producer, distributing 'beta builds' for users to test, react to, and improve?

— ... establishes a profile or brand by showing concern about needs tangential to the real offering (providing parking lots, children's care, queue busting...?)?

— ... creates market, auction, and exchange spaces?

— ... serves to educate customers?

— ... links to partners offering something — a gift — when our coupon is presented?

— ... gives away promotional material with intrinsic value (a tune, an artsy ad, a toy?)?

— ... offers personalization (like the KOR stone in the water bottle)?

— ... involves the family, the bridge club, the hockey team, the kids' friends' parents, friends' friends?

— ... rewards referrals (with ingenious rewards)?

— ... reduces ecological footprint, energy consumption?

— ... distributes rewards and/or awards for ecologically sound contributions?

—   ... inspires users to make it a part of an attractive ritual?

—   ... that interacts with the user in sense-making?

—   ... provides a delightful experience?

—   ... makes the user feel invigorated, positively charged, empowered, transformed?

Ponder cars: speed and performance, for negotiating difficult terrain and hard weather, have been design parameters, with a bit of status and owner persona thrown in. But performance is severely out of sync with how cars actually are — or can be — utilized: we almost never approach the top speed, and spend a large part of the time at stand-stills or moving slowly and haphazardly.

Two American designers, Mike and Maaike,* have suggested that we move from driving cars to *riding*, designing an experience that is social and takes away the tedium, relying upon available technology, like robotics, computers, and GPS. They played around with photos of cars, and found that just altering the proportions of existing vehicles helped create new archetypes. Covering the wheels looked "electric," lowering the panels implied "train," tall windows were "bus." But it was the whole experience that was in need of redefining: *eliminate* driving! Their dream vehicle — labeled ATNMBL — would be designed from the inside out, with elements influenced by architecture, offering living comfort, views, opportunities for conversations, and social connectedness. Scenario:

Summoned by phone, an ATNMBL arrives. One enters from the curb through an electric glass sliding door of standing-height. The entering

*passenger is greeted with a simple question: "Where can I take you?"
There is no steering wheel, brake pedal, or driver — or driver's seat.*

*The interior is like a typical living room, with couch, side chair,
and table. Floor-to-ceiling panorama windows on both sides offer splendid
views.*

*The flat display for the greeting also offers live trip information,
maps, and entertainment. The display slides up to reveal a bar... a driver-
less vehicle allows for drinking... and shuts nobody — young, disabled,
elderly — out. The concept lacks any references to autos of the past:
there are no large wheels, fluid forms, aggressive stance, character lines,
or shiny trim.*

*The ATNMBL's components are densely packed and simplified, pro-
viding drastically more space in a vehicle shorter than most of today's cars.
Electric motors in each wheel, like in the MIT CityCar, provide all-wheel
drive. Batteries are placed under the seats and the floor, with additional
input from roof solar panels.*

*http://www.mikeandmaaike.com/atnmbl.html#.

Josephine Green of Philips Design

As technology merges with our walls, floors, and clothes, we no
longer 'consume' technology, but live with it side-by-side as it sup-
ports and facilitates our daily living, an invisible helper at the ready.
Through this more intimate co-existence, our identity becomes less
about needs ('what do I want?') and more about activity and experience ('how can I best take advantage of what I want to do in the

way I want to do it?'). These take place in the specific *context* of a home, car, public space, hospital, school, or geographical area. In this context economy, passive consumers become active producers of their own lives, as they search for and appreciate ways of interacting with, controlling and creating their environments. People co-create their own content and experience and value anything that enables personalization and creativity. How we light up an environment, monitor a health condition, or access some information will be up to us. Through speech, gesture, or touch we can interface with a display, a hand-held device, or an intelligent wearable garment. What becomes increasingly important is the quality and choice of the interaction. It is about a deep customization, based on a live-in relationship with technology, in which interaction and access are important: an ecosystem of information, services, experiences and solutions.

Our — Philips' — research around value in terms of deep customization has led us to a design strategy of 'open tools'. *This strategy shifts design away from delivering a finished product or experience towards designing an 'unfinished' or 'open' solution that can be completed and evolved by the user or users.* As part of our Design Research program, we have created a number of 'Design Probes' or experience demonstrators to explore this territory together with stakeholders. One example is Nebula (from Philips Design's Noah's Ark project, 2000), in which we moved from designing a new alarm clock to asking the question 'how do people like to experience waking up in the morning?' This led us to develop a simple LCD projector into which the user(s) can download images from the web. Two dots traveling across the ceiling at night meet at the set wake up time, and trigger the projection of images onto the ceiling. Such an 'open' tool allows the users to decide what images they want and to create their own experience. Another such experience demonstrator was developed on the European sixth framework research program, Living Memory (LIME, 1997–2000). An interactive 'coffee table' allows users to intuitively access information about their physical community and to simply take out and put in information according to their interests, etc. In these and other cases, probes allow us to explore new territory in terms of interfaces, interaction paradigms, materials, and systems interactions.

Philips' vision of ambient experience is about this relational co-existence, and by changing the paradigm between people and technology, it has the potential to take us beyond consumption as it has been classically understood. In the context economy, value is generated less through the selling and buying of goods and more through an ecosystem of information, services, experiences and solutions. What we value, rather than what we consume, becomes the issue.

Co-existing with technology in a more intimate way also opens up a more holistic approach to living. Stand-alone products give way to connected and networked environments that enable a more systems-based delivery of value, for example, in the area of health. Ambient technologies can enable a home-centered health system in which users are connected to different circles of care, from family/friends to support professionals, expert patients, doctors, and the hospital. This more decentralized and user-centered system enables people to intuitively monitor and be aware of their health and well-being, whether this is in terms of prevention or disease management. In our research we have explored at how, following a heart attack, patients can gain a peace of mind and quality of life through the unobtrusive and constant monitoring of their conditions. This could take place, for example, by means of an intelligent vest that relays any discrepancies to the appropriate medical professionals. Such systems reframe the question from 'how can we bring people to healthcare' to 'how can we bring healthcare to people'? The same question could be asked, for instance, of education.

Let us dwell on innovation as something — an idea or a family of ideas — that has become successful. Most often, that success has been achieved in the marketplace, but success may take on other faces as well. What about culture, and to what extent would culture be amenable to design? Let us consider some examples.

If you were to visit a Catalan village or town festivity, celebrating something, Midsummer, for example, you just might be struck by a seemingly odd scenery: people dancing around a pile of ladies' handbags, with some small kids topping the pile. The circle dance is *sardana*, possibly (if somewhat doubtfully) with roots in ancient Greece. Variations have been danced for centuries in Catalonia — been danced and been forbidden — the last time by Franco who judged it to be a symbol of Catalan nationhood.

The modern *sardana*, however, the 'long' one, is a more recent invention, music-wise by Josep 'Pep' Ventura, and step-wise by Miquel Pardàs. Ventura equipped the, orchestra, the *cobla*, with eleven musicians — previously there had been a trio — and with a newly invented wind instrument, the *tenor oboe (tenora)*, possibly inspired by him and, in any case, developed by Andreu Touron in northern Catalan (thus in France) Perpignan; these developments occurred in the mid-19th century. The *sardana* thus designed became one of multiple symbols of Catalan culture during the *renaixença* — the renaissance or reawakening — which emerged at this time and grew increasingly vivid.

In the Estonian archipelago, a farm that previously had been a collective has remade itself (redesigned its 'brand') into a farm with Swedish roots — Estonia was part of Sweden for some 160 years, and already in the Viking age Swedish language farmers settled on some of the islands. For the collective that no longer was, this identity or brand had, at best, shaky foundations but after the Russian and communist epoch, a positive identity instead of a threatening one was eagerly sought.

About the same time as the Catalan renaissance, there was a movement in Norway to become more 'Norwegian', politically independent of Sweden, linguistically substituting the written language, Bokmål, which in fact was written Danish, with something indigenous. One leading enthusiast was Ivar Aasen who toured the country to discover or compile the 'true' Norwegian, which he synthesized (designed?) into neo-Norwegian or, in Norwegian, Nynorsk. In reality, no outright substitution took place so Bokmål and Nynorsk now coexist. Several more 'modern' languages have a similar story of being 'invented' or 'reawakened' or synthesized (or, indeed, designed), Hebrew and Turkish among them. Esperanto is, of course, an example, of several, of a fully synthetic — invented — language. (The bare-bones sketches here skip over intricacies and nuances ofa vastly more complex reality.)

Americans, more than others, are alert to how original their political union of 1776 was: to obtain liberation from a 'mother' country — doing so by forming an association between thirteen separate colonies some of them starkly different — different with respect to, e.g., religious creed and economy — and with strong state powers, to create a republic in a world dominated by kings and emperors — Switzerland, the obvious exception — indeed something sufficiently original to qualify as an innovation of sorts (a design, someone?). But what sort?

With the definition of innovation as an idea that has achieved success, the disparate examples given do qualify as innovations, don't, they? But what should the category be termed? Cultural innovation would seem to apply more to new styles like rock, rap, surrealism, or cubism. Social innovation is a category already existing, and for something a bit different. Symbolic or semantic innovation?

Some thirty years ago, two Swedish researchers, Edquist and Edqvist, highlighted the sometime importance of *social carriers* for innovation.[1] Esperanto and Nynorsk obviously crave — serve as carriers for the need of — wordbooks and grammars. *Sardana* is performed outdoors so the *cobla's* instruments have to produce sound sufficiently loud. And the 'long' *Sardana* became a carrier for Andreu Touron's invention, the *tenora oboe*, thus transforming it into an innovation, certainly a designed one. For unknown entities, mathematicians use x, y, and z. Let's choose to apply y and term this kind of innovation *ynnovation*.

Cause-and-effect relationships can often be disputed or discussed, possibly generating an understanding of something as more of a symbiotic relationship. The driving force behind all the examples here seem to be a striving for identity, and that identity was not just the *sardana*, Nynorsk, Turkish, or the American constitution but something wider and more profound. What other forces might be conducive to ynnovation, with its elements — which are they? — of design? What might be the carrier for or impetus to ynnovation, and what might a particular ynnovation serve as a carrier for? What technologies and innovations might be spawned or brought into existence symbiotically, by or with ynnovation? Facebook and social media, in general, come to mind since they are currently so topical. As always, the challenge is to perceive the coming topicality before it occupies main stage.

So, our scope should be widened to encompass *ynnovation* as well. The attempted contribution here is to give it a name, to design it in the sense of giving it a sign without which it might go undiscovered and untapped. Indeed, it is a further opportunity for the exploration of a foreign ecology.

[1]Edquist, Charles & Edquist, Olle, 1979. Social carries of techniques for development. *Journal of Peace Research* 4, Vol XVI 313–331.

Endnotes

[1]http://www.foranet.dk/Publikationer/Rapporter/Data/Concept%20Design.aspx.
[2]Cf. Chapter 10.
[3]Senge, Peter M, 1990. *The Fifth Discipline: The Art & Practice of the Learning Organization*. Doubleday Currency, NY, NY.
[4]Forlizzi, Jodi, 2007. The product ecology: Understanding social product use and supporting design culture. *International Journal of Design,* 2(1), 11–20.
[5]http://www.psfk.com/2009/08/good-brands-report-2009.html.

Chapter 8

COMMUNICATING WITH AN IMAGINARY DESIGN

Because we are considering, *thinking* design, and as we do so, striving for innovation, we venture into unknown territories, to discover and profit from new opportunities. There can be no detailed plan for how a final solution will turn out, nor any preordained process or procedure for arriving at it. It is a quest of exploration, of experimentation. And we do know that *to nail down what is thought to be the final design, to choose a direction* **too early** *into a project risks leaving one ending up in a blind alley, losing time and resources when early choice was thought to save them.* Experimentation is the order of the day, which equates with frequent prototyping and the pursuance of several options at once. Wasteful? Rather, compete with ideas internally than have some competitor in the market beat us to the eventual successful design! Frequent prototypes and many experiments also mean frequent opportunities for learning, a sort of learning that will benefit the project generally as well as spill over into other projects, immediately as well as later.

Think sketching! To sketch adds an important dimension to ideation and innovation activities, an effort where the key is an ability to illustrate potential novelties before they exist at all, before deciding to

205

make them real, at first rather generating a better understanding of what they would mean. Sketching is fluid, it is fast, and has a redeeming quality of essence: swift, frequent prototyping. Yes, think sketching, but consider also sketching as an analog, the function of sketching transferred or translated into other domains than necessarily that of something drawn by hand or even being visual. Certainly, thinking visually holds out the promise to avoid verbal overshadowing, but what other ways of 'sketching' might there be? Furthermore, there is a paradox in that too accurate a rendering may turn out to be counterproductive — or so is accuracy beneficial; the one or the other depending upon objective, depending upon ideation or development stage, and depending upon who is the one pondering the sketch — different individuals communicate and interpret differently, verbally, visually, with gestures and body language. Invent and explore new languages — new types of prototyping and sketching, exotic but versatile materials, powerful computer software...

Sketching goes to the core of ideas and solutions, it highlights the essentials; it simplifies since that is the essence of sketching. The vagueness and imprecision, the mistakes and uncertainties in an early sketch may serve as triggers to creativity and provoke interesting associations. Conversely, a seemingly finalized CAD (computer aided design) rendering of a product just contemplated may cause the closing of the mind prematurely; that is why accuracy may be detrimental in the beginning of an exploratory process. Later on, however, the possibility to arrive at a very precise and seemingly finished reproduction of something existing only as a data set may serve well in convincing investors, customers, and other potential stakeholders that this may, in fact, be a promising product.

Just as an aside, frequent prototyping has, particularly for computer software, spawned a new development process, termed 'agile development', which is also tested in other design domains. This is characterized as being an alternative to 'the waterfall method' of repeated feedback loops, since it is, instead, focusing on frequent prototyping, each 'build' relied upon as a stepping-stone to the next. Projects are subdivided into limited tasks distributed to several teams working in parallel, progressing through successive iterations.

Instead of 'think sketching', we might have said 'think scenarios' and also 'think playing': scenarios are not necessarily just verbal descriptions, and play, as in *play around*, applies to a broad array of activities. On a more general level, we might say 'think representations' which, apart from models, mockups, and prototypes, includes evoking similes, allusions, associations, analogies, allegories, and also metaphors. Gestures and body language are further instances of representations, and, as with sketching, faculties for employing them can be developed.

So when considering the various ways of representing ideas, designs, products, services, experiences, or emotions, always ponder whether the type of representation chosen might risk excluding or hiding some important qualities or dimensions, or be lost in certain modes of perception. Furthermore, ponder whether it opens up vistas to the spawning of new ideas, whether it challenges the mind, whether it inspires to design and to innovate. Visual thinking, for instance, activates different parts of the brain from verbal thinking, but then think about music, haptic experiences, odors… Images may often communicate much faster and better than words — the latter may even be detrimental (recall the verbal overshadowing effect). Images can transcend language and discipline boundaries. Mathematical and geometrical thinking would neither easily fall into the visual category, nor into the verbal one.

Sketching by hand obviously applies to physical objects but also to people and places, thus, for example, to user situations, interactions, and interfaces. It is one way of visualizing physical spaces, of making scenarios vivid, and to capture an activity on a storyboard. But it may also be applied productively to more abstract ideas and concepts, to illustrate their core qualities, structures, and interdependencies. Here, sketches contain systems models, flow charts, diagrams, figures,[1] possibly combined with inserts of physical object illustrations. Complex relationships and interactions may thus be reduced to good overviews and striking insights. For a particular project, such sketches, produced at an early stage and then with fresh ones produced regularly during the evolution of the project, may contribute to an evolving common frame of reference and a project language. Such communication may serve as an important shorthand way of communicating within the project team although team members should be aware that it might not be as easily accessible for outsiders. With this caveat in mind, dialog with stakeholders outside the project team may still be much furthered through creating sketches for overview and insights.

As an aside, we would like to mention that sketches may also serve to record meetings and structure relationships, for example, with mind maps.

All kinds of models, mockups, and prototypes, perhaps with rapid prototyping machinery, such as in Styrofoam, wood, papier mâché, or clay, could also be regarded as sketching in a general way — certainly as visualization. The computer-based methods can provide three-dimensional renderings, rotating or moving the 'sketch' in realistic ways on the screen, perhaps even in space, allowing a product to be assembled and disassembled in a virtual world. Thus, end users, investors, engineers involved in manufacturing may observe and react to amazingly realistic representations of products still at the idea stage. How might this contribute beneficially for you? Do you feel a need to sell before producing, to raise capital for products just in CAD format, to discuss intricate details with manufacturing people along your value chain?

As a Swedish designer stated, 'computers are not tactile', so many designers prefer hands-on methods for producing models, mockups, and prototypes emulating the sensory experience of working with traditional materials; traditional tools are familiar and so is the 'feel' of the material and its feedback. When relying upon computers and other equipment, interfaces must be direct, intuitive, and user-friendly, easy to learn to use, but tool versatility is essential too, including comfortable input methods. Forms must be arrived at in correct proportions and shapes, without any irritating or constraining restrictions. The more details can be made precise and attributes added correctly, the better. Surface quality as such — the 'feel' — may also be a factor to consider.

With Archi-Me, the UK company MOOFU provides a system for transforming 3D CAD models into fully interactive three-dimensional (3D) environments. The 3D virtual world thus created can be explored and changed by people who employ a 3D character, an avatar, to look around within and outside the various sides of a new design. Several 'representatives' — avatars — may be present simultaneously, discussing and rearranging and testing the configuration of objects or details of a design. Models can be exported as software or embedded in a web site for use of its visitors. Z Corp, a spin-off from the MIT, is a company that offers a whole range of 3D "printers," several of which allow color objects to be produced (Fig. 8.1). The same company also provides a 3D scanner for the input from a 3D object, sculptured perhaps in clay or styrofoam. There is a host of computer software available for the generation of 3D representations from CAD data,[2] while Zebra Imaging[3] is an example of a company that offers something extra — *holographic* true 3D images produced from CAD files.

Rather prototyping than sketching, new ideas and designs may well be tried out, with numerous participants and contributors, in virtual worlds like Second Life. Virtual estate sold there can be very valuable, not just in Linden™ dollars (the currency issued by the company running Second Life) but in real ones. Likewise with designs.

In one particular electronics hardware project, the design team created hundreds of sketches and more than 150 foam models, exploring a broad range of forms and features. Many alternative sketches are

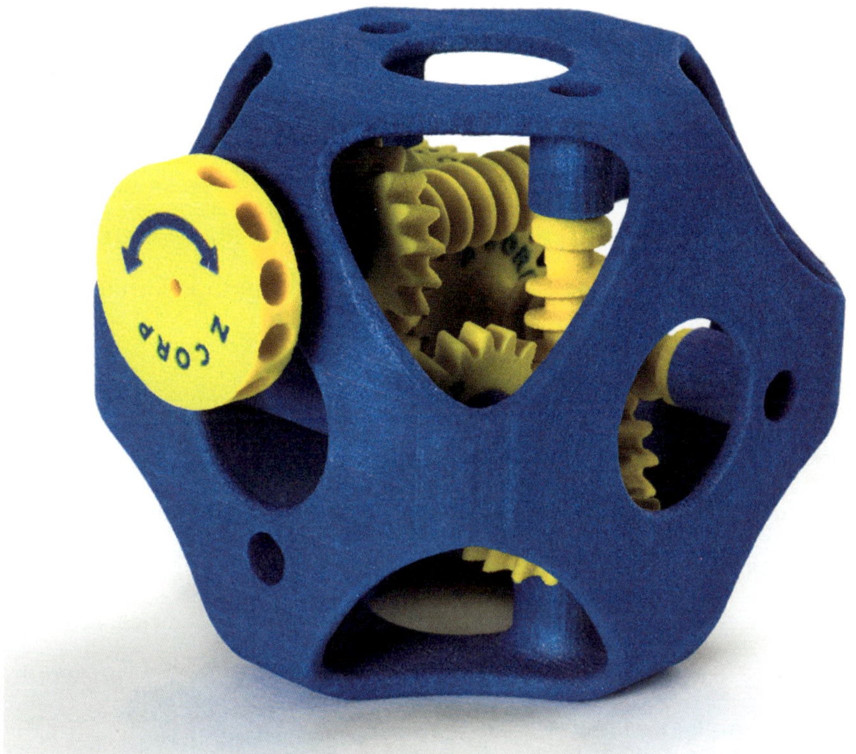

Figure 8.1.

demanded early in the development of a new car — typically, 15–20 — to allow for the discovery and exploration of different scenarios as well as of different design languages or language expressions.

Sketches, whether by hand, computerized, or carved out of plastic foam, may be located to user environments and, here, also related to customer segments. One of the many ways to do this is by envisaging those personas, "synthetic customer personalities," which are seen and described as real. In a way, they are personified user scenarios, useful beyond the design context, as we have seen.

If a sketch, whatever its form, gives the end user an idea about functionality, it may, in addition, offer more than a hint of the design language applied. A series of sketches may offer more about what kind of "soul" or what meaning the product will have, or tell more than any well-phrased verbal description; emotions, messages, and meanings are tacit, not so easily described in full with words, formula, or mathematics. Sometimes, they are impossible to describe.

The very fact that sketching is a means to convey and receive messages about something that is tacit is of major importance. Without it, a user may not be able to express her demands or know what she really prefers. 3D visualization has turned out to be a good facilitator for discovering tacit preferences, knowledge, or competencies; it assists in laying bare what words cannot fully pronounce. Dealing in words, market studies and polls have to leave out questions about tacit issues. By contrast, focus groups conducted by individuals with keen eyes and ears for vague, transient psychological signals may well give some clues by picking up subtle signals of tacit feelings — have you acted so as to establish the availability of such competency? One might claim that sketches offer an interface between the tacit and the verbally expressed.

Passenger cars are prime examples of products designed to make statements, that tell stories, about their owners. Given that fact, and that the greatest opportunities for affecting the outcome of a product development process exist at the beginning of the process, having the end result available early on, in sketch form, might have a great positive impact. The advent of computer assistance has greatly reduced reliance on the sometimes cumbersome — expensive and time-consuming — production of mockups and prototypes.

Several media may be required or productively combined to visualize something that is still on the drawing table, or in the computer memory, depending upon the stakeholders involved, the very problem, and a number of other factors: contingencies. One is the shear power to convince or to communicate accurately and powerfully; models may surprise and create awe.

"Prototype" is a word seemingly associated with the physical world. For more abstract interrelationships, we may resort to models with a higher degree of abstraction that can be employed for simulations. While a scenario is a composite view of a situation, often of the future, a simulation depends upon a model, those abstract relationships, which are relied upon to make a picture of how the future will unfold — typically a quantitative scenario, or a family of scenarios, of simulations. If simulations allow for more people to be engaged and active — all the better. All the better, also, if simulation models are open for tweaking, that is, discussions about relationships, measurements, goals, etc.

While not produced quite as rapidly as sketches, prototypes may still equate with gains in development time since they resonate more with end users. It is much easier to realize and articulate what you want through trying and testing prototypes than by establishing a list of requirements. End users may constitute a key group, but there are others, such as team members and others associated more or less closely with a development project. Here, again, sketches, mockups, prototypes are powerful communications devices; after a time, they have become part of the evolving language of the project, functioning as points of reference, symbols, and stepping stones within an ongoing dialog. In a time with much talk of social media such as My Space and Facebook, prototypes may be seen as social media also — or community building platforms. We have mentioned the importance and the power of *boundary objects*: representations of the core object of a project.

Designing with a desktop computer: what does the particular input tool mean for the actual design work? Does it affect the speed with which sketches can be produced; does it give a sensation to the 'sketching' hand that it is comfortable enough, translating intentions truthfully? Demands may vary between aircraft, pleasure boats, Web pages, office chairs, children's toys, affecting the choice between, for example, a mouse, a sketchpad, a joystick, steering wheels, a Wii controller, also a sponge or a particular type of 'clay' which may be manipulated and given various forms in 3D, all alterations recorded digitally in the computer. The 'clay' or 'sponge' functions may be seen as opposite to a CAD process in the sense that what is formed by the human hand is translated directly into computer data (cf Z Corp's 3D scanner mentioned above). Thus, these materials are flexible and may be deformed, like clay can, and unlike rigid Styrofoam or wood. The Wii controller type of input, which for its functioning draws essentially upon an accelerometer, would allow for input to be produced by way of body language and gestures.

Haptic devices are those that use touch and that are tactile. In more advanced forms of virtual reality, gloves or even an entire body-stocking may be relied upon to create a sense of mechanical touch, the next challenge being to develop it to feature texture, temperature, softness, vibrations, humidity, and the possibility to evoke pain, the sensation of (im)balance, or the feeling of being tickled. Rapid prototyping allows for 3D prototypes to be created and may also function like '3D faxes' in the sense that computer instructions can be transferred from the location where they were generated to other places, for 'printing' at these other locations, possibly on the other side of the globe. The challenge, however, in creating a *real* virtual 3D space feeling is to reproduce the tactile sense — yes, a tough task. Holography allows for visual 3D representations without the need for spectacles. It is worth noting that it was reported in 2009 that a team of scientists from Italy and Sweden had developed what was believed to be the first artificial hand with feeling, and with tactility. It was attached to the arm of a man who lost his own hand through cancer and it works by connecting to human nerve endings with tiny electronic sensors.

A storyboard is a comic strip-like rendering of a series of events, a narration of a story showing interactions between what is designed and its users and others affected. Like comics, it is often hand drawn but may also consist of photographs or computer renderings. Like in any scenario, there may be branching into several alternative storylines, or perhaps for different segments of users; like with our sketches and prototypes, different variations and alternative outcomes from uncertainty may be depicted. Just like in a scenario, the story must be coherent, reliable, and proffer validity. Basically, while it may ask for deferring disbelief it must be internally consistent. There are storyboards in 3D, and there is computer software available for storyboarding.

As we have seen, for the development of the Vigix product dispenser, Alvarez and design consultants at IDEO told a story, and built a user scenario. A scenario is an outline of a drama or a synopsis for, say, a movie but is, perhaps, most often used to depict a 'synthetic future,' that is, a picture of a possible future, maybe also with the road leading there unfolding. It is emphatically *not* a forecast, and to demonstrate various options, uncertainties, and pivotal events, several alternative scenarios may be constructed for the 'same' future, for a region, a country, a product category, and so on. As mentioned, it should be internally consistent. Depending upon format, it may be qualitative and quantitative to various degrees, variously fine-grained or just roughly outlined.

Employed for designing, the synthetic future quality makes for placing the object of design — product or service — within its larger context of application, possibly including the access road, that is, market dynamics upon market launch. Since this is about communicating with the future even more than sketching or prototyping — prototyping the future — choice of medium and presentation and especially of striking details, possibly metaphors, are important. Even a written scenario may be designed in several ways, such as in the form of a putative newspaper piece or a corporate annual report; a PowerPoint presentation is another possibility, a movie yet another. Or what about a theater play — design team members taking on and interchanging different roles as the performance is repeated. An exhibition or a future environment, such as a home or an office, are other possibilities.

One might see sketches of potential future products or user situations as scenarios that offer a more complete user environment — the product sketches serve as crude scenarios, since they come without much of any context, which is why 'environment' may be important to add. These dynamic representations can provide a sense of completeness and suggest how the product or service might operate in conjunction with other devices or services. As with sketches, scenarios are tools for communication, allowing for discussions of and reactions to a product only envisaged. Scenarios may be seen as constituting careful compilations of sketches.

Analogies and metaphors represent sketching — obviously, the noun 'sketch' is employed as a general analogy in this chapter — in a more abstract sense, taking concepts from one world to apply them to another, like sketches are simplified and 2D or Styrofoam or wood made to represent metal or perhaps a whole ensemble of components and, thus, all the materials making up a product. Since analogies or metaphors bring concepts, features, facts, or phenomena from one world to another, there is never a one-to-one representation. So, again, there is something like that imperfect, challenging and inspiring imperfect sketch, a challenging, positively bothering, discrepancy besides an enlightening parallelism.

A metaphor works through transferring associations from one experience to another, serving as associative shorthand to help people understand something new and possibly alien. Thus, metaphors may assist those involved in design to get a handle on unfamiliar design problems, juxtaposing them with known situations, and they may allow or force the design team to think unconventionally and foster the application of novel ideas. Even though we are not always aware of it, metaphors develop out of the way we handle our body's interactions with its environment and affect how we perceive the world, how we categorize experiences, and also how we organize our thoughts. Sketching and prototyping, relying upon metaphor, would be trying and testing several metaphors and thus *experimenting with metaphors.*

Metaphors, in particular, hold the potential of helping getting to grips with ill-defined problems. Reasoning in metaphors iteratively may allow the design team to successively enlarge its knowledge of the larger design context, a new unique one which has been fleshed out, relying upon metaphors in structuring it. Fuzzy metaphors, and there are many of them, may assist in reflecting on the core of the problem and its context, which perhaps are inseparable, at this stage.

Analogy is a profound similarity while metaphor is a pictorial one; depending upon definition, metaphor may be seen as a category of analogy. An expression like 'the iPod ecology' comes with a whole set of parallels between the various features, objects, services, and offerings associated with the iPod, and an ecological system in nature. The expression 'the iPod economy' is another analogy, not far from the first one, but evidently not identical since it is oriented more to the economy and the market, ecology more towards evolution and interaction (co-evolution also) in nature. An analogy rests upon a relationship, and since that relationship is not an identity, several different relationships may be drawn upon to generate several different analogies, depending upon what relationship and what particular quality or dimension is highlighted. When bikes were becoming equipped with small engines, they were called 'bikes with auxiliary engines' but also 'light-weight motorbikes', sometimes depending upon the regulations applying to this category of vehicle in a country. Later, in some places, they were called 'mopeds'; in Denmark, 'knallerter,' evoking their sound. Cars were called 'horseless wagons' early on but the word 'auto' stems from another term, going back to 'self-propelled' or 'autonomous motion.' Again, analogies offer a laboratory for experimentation, and analogies as well as metaphors may be instrumental in defining new design languages and new meanings embedded in the design.

The next level of abstraction is working with symbols. This might seem to represent too long a stretch but we should remember that numbers are some of our most frequent symbols. And there are many specialist fields of design, including graphic and Web design.

The power of visualization:

	Total bandwidth (bit/s)	Conscious bandwidth (bit/s)
Eyes	10,000,000	40
Ears	100,000	30
Skin	1,000,000	5
Taste	1,000	1
Smell	100,000	1

The wine-lover — whom we have met previously — knows well that she is dependent upon a set of similes, metaphors, or analogies when it comes to describe the nature of a wine: it comes 'full-bodied,' 'astringent,' 'with the bouquet of red berries,' 'peppery,' etc. The gastronomist would, at least, depend upon spices for the characterization of a meal. But outside the fields of designer wines and haute cuisine — Ferran Adrià and elBulli with molecular cuisine come to mind — how may taste and odors be harnessed in other fields of design, in industrial design? Metaphorically, yes, but more directly? Likewise, the design of sound, not just composing music: cars and motorbikes are products directly associated with sound sensations — recall the Danish 'knallert' — one might say sound signatures, such as when Harley Davidson were granted a patent for a particular sound. Similarly, artist Yves Klein has patented[4] *his* color blue, *International Klein Blue.*

You have met a number of implicit suggestions in this chapter. To spell a few of them out expressly:

- Define what contingencies, or what qualities define, or should or might define, your business, and what implications they will have on methods and equipment for sketching, prototyping, etc.

- How may sketching and prototyping be made swift and frequent? What about moving from 2D to 3D and even 4D, that is, incorporating the time dimension?

- What are your communications needs; whom do they involve among design and development team members, investors, suppliers, marketing people, employees, customers?

- To what extent and in which forms should you consider relying upon story-telling, storyboarding, and scenario production?

- How may sketching, prototyping, the employment of analogies and metaphors contribute to a better understanding of tacit, silent knowledge, perceptions, and preferences among customers but also within project teams?

- At some point, the design has been sufficiently crystallized so as to be embodied in a mockup, while much of the inner workings still are subject to considerations. It is of particular importance to create and highlight a physical representation, solidifying problem definition and focusing discussion and also easing collaboration and coordination. Actually, there is a statistically significant correlation between the existence of such a spotlighted representation and project success.

- Some organizations have established a visualization center to great effect. Depending upon activity, you might consider establishing such a center, or obtaining access to one. Perhaps benchmark one or some that operate? Perhaps establish one in conjunction with a university?

In search of failure — designing for failure?

Who would be searching for failure? Failures just happen, though we always hope to avoid them. A vile thing, a vile word, no?

On the most aggregate level, failure is, in a sense, a consequence of success: no innovations, no business models live for ever — there is always Schumpeter's gale of creative destruction. Failure is trivial: nothing lasts for ever. General Motors once held 50 percent of the US auto market; later (much later) they had to file for bankruptcy protection. TWA and Pan Am were once dominant airlines, now resurfacing in old movies only. And so on.

The best we can hope for, no, the very best we must put our trust in, is the existence of provisional truths, provisional, ephemeral successes, to be enjoyed as long as they last. It is not that technology moves on, though it does, but society changes as the effect of an intricate interplay between new ways of organizing, new conventions, tastes, and demographics. Novelty and disruption is the order of the day, not the economist's fake equilibrium. All winners eventually become losers, so move on. Everything is a failure in the long term; thus, nothing is everlasting — never completely.

Again and again, firms and organizations that display surprising revival do so because they have succeeded in finding, within the depths of their structure, some odd project that is married to a technology or a market with a booming future. More often than not, that odd project was killed off on several occasions, or survived just because it was protected by some maverick with sufficient clout, or because it was simply hidden or forgotten. A *something* identified as a failure but a failure that, for some reason, the organization *failed* to kill.

In a sense, every success is built upon an assembly of failures. The challenge would be to detect them; memory would tend to edit them away.

I am in search for a better word for failure, one not tainted with negative connotations. Because, after all, failure is the high road to innovation, because failure is the inevitable investment in the future, because failure is — the road to success. Remember Edison retorting, when pitied for having failed at 5000 attempts (or some such number) to get the right filament in his light bulb: "no, not 5000 failures but learning 5000 ways that won't work."

Learning, tests, trials, and experimentation offer alternate, more positive angles. Ashkan Pouya and Saeid Ezmaeilzadeh, innovation practitioners-cum-students, talk of 'the creative dance.' As kids, we experiment constantly with language, with acquiring skills such as biking and swimming, maturing, going through a process of socialization. Learning to ride a bike is worth some bloody knees; corrected grammatical errors are just that — but no failures. Yes, we talk about learning and maturing, not about failing. To the contrary: we are succeeding!

The title "In search of…" intends to invoke that business bestseller "In search of excellence." I heard its (co-)author Tom Peters tell what secret they had found behind the fact that some oil drilling companies were, by far, more successful than others — very simple: "they drill more holes"! The more failures, the larger the number of successes. Losers become winners.

Within hailed innovator 3M, seasoned chemist (with more than 20 patents to his name) Spencer Silver happened to develop a glue that did not hold. He suggested several applications but none took off; he also told of his invention at several internal seminars, one attended by Art Fry, choral singer in his spare time. Well, you know the rest: Fry took Silver's paper with bad glue to serve as bookmarks in his hymnal, and the Post-It® was born. In effect, Spencer Silver had often told of his 'failure' so that others could learn from it, and build on it.

So, should we not fail often, fail quickly, fail inexpensively. Tell and disseminate failure results overtly — do not cover them up or try to forget them — and attempt to learn all lessons embedded. Look upon failure results as weighty trade secrets.

What about designing for failure: design strategies for generating complementary failures that, together, make for success? What about rewarding failure — yes, there are firms with rewards for "risks that were worth taking even though the ideas eventually did not pan out." So, think about metrics, about failure inventories and audits, always with a view to profit from the learning experiences. Map out a failure strategy. What failures would generate most information, or the most useful information? What failures would be complementary rather than overlapping? What failures would be best for us, least worthwhile for the competition?

Yes, we need to find and design and practice ways precisely to fail often and with low costs. To re-use materials and what otherwise would be waste. To rely upon scenarios and stories, and not least upon sketching. To make rough, speedy models, mockups, and prototypes — using whatever proxies are easily and cheaply available. What more? How to map the best set of failures to set out to make, to have the aggregate results contribute positively in the end? What about failure experiencing as part and parcel of every individual's development, of an organization's establishment of a joint culture and sense of mission and cohesion?

We — you — might need a taxonomy for failure. One type of failure might be the one associated with extrapolating something already existing, testing the boundaries as it were. Another type of failure would be one implying testing something never tried before, an exploration of the unknown.

This text box is a failure — no, it is not yet complete (will it ever be?). A try, an attempt, a test, an experiment. So, why not help me out?

Endnotes

[1]A fantastic map of different types of pictograms is found at http://www.visual-literacy.org/periodic_table/periodic_table.html.

[2]This site contains a number of free 3D tools: http://www.optionprophet.com/blog/.

[3]http://www.zebraimaging.com/index.html; here is a Youtube video of how it may work: http://www.youtube.com/watch?v=jx3TSQul94E&feature=player_embedded.

[4]French patent no. 63471, dated May 19, 1960, a patent for the chemical composition of the pigment Klein developed.

Chapter 9

BUILDING BLOCKS TO DESIGN

Before Henry Ford, any passenger car was unique: even if two cars shared the same name mark and model number and looked similar, parts from one would not fit the other because of the fact that each and every part was crafted individually for that particular car. So one of Ford's most important novelties was the introduction of a modular system, residing upon standardization; together with the Swede C E Johansson, who provided the measuring system, he led the standardization of the exact measure of an inch. Clearly, this Ford system was a condition for mass production and the resulting drastic reduction in the cost for a car.

For cars, that popular design item, there is an entire dictionary of styling and engineering terms; a brief glimpse will have to be sufficient. A platform is a chassis but much more; it includes components or subsystems such as power-trains and suspensions (cf. below). The term 'screen angle' refers to the angle the windshield deviates from the vertical. In the late 1950s, American car manufacturers introduced 'panorama windows,' Ford's and GM's ending vertically, the Chrysler Group's at an angle. 'Surface expression' can be, e.g., flowing, soft, and organic or, conversely, boxy and geometric; it comprehends characteristic brand elements such as grills and lamps.

The availability of components, subsystems, and standards obviously has important ramifications for design. But their existence does not make design choices superfluous: these stay essential and non-trivial — critical, even. What about designing a car, a music system, or a computer with the "best" components available — would the result not be a supercar? No, not at all: in the case of the car, it would probably not even be possible to assemble them into a functioning vehicle. As for the other cases, it would be difficult or impossible to choose the 'best' components without knowing about some of the demands to be met.

The starting point cannot be the components, selected for singular component qualities, but should, rather, be the system, establishing a concept for the whole. So an overall concern has to do with what is the product or system *architecture*. The architecture may be given by technology or the result of a design choice: between them, front wheel drive cars may share the same architecture, certainly different from that of four-wheel drive cars. The MIT City Car project in our portal chapter of case stories features a radically different architecture, with electrical motors integrated into the wheels, the wheels being omni-directional, and the cars partly collapsible.

If we, instead, think in product families, qualifying family members should display some commonalities (in some instances, what is here termed 'family' may be given the name 'portfolio' or 'brand:' since cars, furniture, and kitchenware are quite different, the concepts are somewhat fluid and seemingly overlapping). There may be subsystems that fit into several products of the family; there may be a commonality in user interface and interaction rules; there may be a particular family style or design language. From production, warehousing, and logistics points of view, commonalities and standardization among components would be beneficial, and, in addition, they would make for easier servicing.

A product architecture is an abstract outline of the essential sub-systems and their interconnections and interdependencies. A particular product design is a specific instance of the overall architecture. Its closest siblings probably share the same architecture but are variations on it, to be fine-tuned to different feature demands and price levels in the market. For an innovator particularly, launching a novelty would equate with establishing a new architecture. It would be all-important for that architecture and its concomitant design structure to be well thought out. If it were relying upon modularity, this might allow for offering many product variations in rapid succession, to try the market and also to serve its various niches without forbidding expenses or delays. 'Familiarity' in general, re-use of design elements, components and sub-systems, may not just cut costs but also improve quality and ease-of-use.

Another important concept is that of platform, possibly best known from the automotive field. This is a central design component or key sub-system that is recurring in a series of different 'models' or even brands. As the very concept indicates, other necessary components and sub-systems are bolted onto the platform, the would-be metaphor literal, and these may vary between models and brands, giving them different features and performances.

Considering design options, the market must be taken into account and addressed in its entirety. Just focusing on the market corner that allows for higher prices and higher margins for niche products offering maximum performance may be a dangerous trap, leaving it to rivals to offer simpler and less expensive models. They might then, from that base, progress to offer performance to meet the requirements of more demanding customers as well, eventually invading that high-price, high-margin space. Conversely, the prestige and trust generated from a brand and an architecture catering to the high-end market often pays off on more mundane offerings from the same source. One should note that independent designers may be more aware of or more diligent in deciding what segments and critical price points that exist than their clients, as they integrate across a range of clients.

In our previous book *Design-Inspired Innovation*, we illustrated various modes of modularity graphically, adding that there may be many more than those mentioned there. As an example of component sharing, we mentioned a re-chargeable battery shared among various tools such as a drill, a jigsaw, and a sander. Component swapping modularity was illustrated by the example of a drill that may also be equipped with a sanding disc or a grinding head. Cut-to-fit modularity was exemplified with the case of tailored suits. Mix modularity is what we experience when visiting a paint store or a Chinese restaurant, where a broad variety of colors or meals, respectively, can be compiled from a limited stock of pigments or ingredients. Though without noticing it, bus modularity is encountered in our use of computers or telephone systems where all the various components are coordinated by means of a standard communication path, which allows for the simple installation of additional parts. Pre-fabricated building systems display sectional or snap-together modularity; a good example is the Lego block system. How would you describe the modularity of a standard deck of cards, where the same deck might be relied upon for playing a large and, in effect, unlimited number of card games, like poker, bridge, and canasta? The cards provide an unchangeable piece of hardware, the game rules the software that can be re-programmed radically.

So the general point is that by applying a modular approach, the result may be a huge array of combinations, variety, and experimentation. Seemingly a contradiction, standardization and modularization may allow for more of mass customization. In *Design-Inspired Innovation*, we related the example of Matsushita, which offered its Japanese bike buyers the choice between 11 million options, based on a modular system, including an electronic 'catalog' to assist with choosing.

Modularization has repercussions for the organization of design work. The more well-defined modules and sub-systems are, and the more independent they may be, with interfaces to 'their external world' defined by some system-internal or universal standard, the more design tasks may be divided between teams, delegated, or outsourced. An important issue pertaining to innovation is on what level innovation may take place, and also whether it may turn innovative to fashion or to design, those very networks of collaborators. One level of innovation might well be that of the interface between subsystems, whether to share data or energy or have them provided separately and independently to different parts of the whole.

Then, there is the important question about standardization, about the choice between open and closed systems. Standardization has many benefits, such as cost reduction and customer confidence and appeal. But standards may reflect previous generations of technology and thus impede innovation. Creating a closed system and a lock on the market has its own allure — often enough a trap. For modular design to deliver maximum benefit, the interfaces between modules must be transparent and open rather than closed and proprietary: it is like being part of a community — no, it *is* being part of one. Transparent interfaces create the mixed blessing of opening up the supply chain for modules to competition, making the producing firm more vulnerable to outside influences and predators. But at the same time, it as well as its customers will profit from a wider range of choices and an increase in the rate of technology advance and the breadth of potential applications. In most historical cases, a design that allows for more competition, and by the same token, more implicit partners, wins a larger share of sales, eventually achieving dominance.

For us, the terms 'module' and 'term modularity' apply to design and technological systems, though they have broader applications, such as on organizational efforts. A product system or family may be said to be "modular" when it can be decomposed into a number of components that may be mixed and combined in a number of configurations. These components can connect and interact, and also share or exchange resources, like energy and data, relying on the same interface. The interchangeability of parts, modules, sets this apart from a unitary, tightly integrated product where almost everything is set to work exclusively in the given combination. However, cars of the pre-Ford type are now rare; only extreme sports cars are manufactured that way, that is, crafted individually, as is handicraft.

Modularization may make for designing at different levels — system versus components or sub-systems — which may save time and cost, and allow for the 'recycling' of design elements. The upside of cost and time savings, easier servicing and plug-ins offering capability extension to the customer has its corresponding downside in less design freedom and conceivably less streamlining to specific functionality (though we recall the Matsushita bicycle example). With intelligence modules, that is, computer power, more and more integral to a host of products, software modules are becoming more common as uniquely flexible ones, most often available for updating and augmenting.

Just as an example: the Swedish drawbar couplings producing company VBG AB, with the assistance of consultants Modular Management AB, has been through several modularization projects, resulting in vastly improved profitability and reduced lead times, from 21 to three days for one major product. Here, assembly time was reduced by more than half, and the number of unique parts almost by half. The integration of an acquired German company was much facilitated by the modularization approach. The figures noted indicate the utility of metrics, for e.g., important value drivers and levels of simplicity or complexity.

Modularization allows for experimentation on the module level, and for developing alternative modules for the same task. Doing so in parallel allows for employing the real options methods to calculate the relative potential economic outcomes for different alternatives.

There is one design choice that is both hidden and glaringly obvious, and that is the default.[1] We tend to think of this factor in connection with computers but it is found in all technology and even more broadly. Many hand-tools come designed for right-handed people; cars were, for a long time, designed for men; troublingly, the latter still holds for part of the medical world. When changing the batteries in a camera, I may find it flashing 00:00 or perhaps 12:00, whatever the default. The challenge is to make default creation conscious, give it a purpose, not leaving it to chance.

The development of the world's best-selling telephone switching system, AXE from Ericsson, provides a case-in-point for modularization. Previous attempts at designing digital switches had met with tremendous problems in producing software, that tended to be too entangled, 'spaghetti', and with software bugs spreading unaccountably like viruses. The solution was a rigid modularization, of digital circuitry hardware, of software, of data, and also of the mechanical system for organizing modules in racks. Interfaces and communication between modules were standardized, and units were equipped with independent energy sources — which implied a certain cost; not an easy decision. The result was a system that, to customers' surprise, functioned from day one; that was eminently scalable; that allowed for a shift from analog to digital switching; that allowed for the addition of new services; and that made for modules coming into testing and production just as they came out of the design studio (that is, the electronics laboratory).

http://www.ericssonhistory.com/templates/Ericsson/Article.aspx?id=2126&ArticleID=1904&CatID=&epslanguage=EN; Vedin, Bengt-Arne: Teknisk revolt. Atlantis, Stockholm 1991 (in Swedish).

If we take the more mundane interpretation of the word 'design' into account, that which applies to fashion as well as cars, chairs, and camera bodies, we may distinguish a number of 'formal' building blocks. One is shape, which equates with geometry and form, but also this particular geometry and this particular form in relation to the surroundings, the environment, and the very context. In perceptual psychology, this may be called figure and ground, and, thus, it is not just the question of the product but also about its relationships to its context. Here, both the perceptions of customers and the offerings of competitors should be considered.

'A diagonal line draws the viewer into the picture' is an adage from a photographers' handbook, so lines, connecting elements with one another are important, and those 'lines' may have to do rather with a listener's ear or a reader's thinking: a sense of invitation, a sense of evoked motion — into the design. Scale may be used to underscore certain design elements ever so subtly, or not at all subtly, when certain features or design elements are consciously highlighted. Perhaps the translucency of technical objects that Apple popularized and that had a viral effect may be seen as scale transferred metaphorically. Scale is related to proportion: to enlarge is to keep something else small or its natural size, but then, of course, proportions may also be distorted, to achieve certain effects, perhaps to add clarity or user friendliness. Variety and its pacing and rhythm (including repetition) are complementary dimensions, as is contrast, all of them applicable in visual arts as well as in music, working together as composite elements with the others to instill harmony, all of it coming together in structure.

Depth is something which, if we attempt to generalize it, means intriguing, fascinating, not-so-easily-decrypted, alluring, evoking references, staying in memory. Stance has to do with balance and attitude, texture is the feel of a surface, but that feel may be perceived just by looking. It is also about geometry but more of a structural, three-dimensional and tactile one: optics, cold–warm, edges of a surface. Also, colors vary from cold to warm, and they carry other connotations as well. For a market niche of 'traditionalists', for example, Philips developed a product range with an off-white color scheme associating with ceramics, thus alluding to craftsmanship and quality. Other visual as well as tactile references were likewise related to noble materials.

When it comes to form qualities, the several mentioned and somewhat discussed are not even the only ones; though all are justly called designs, there are substantial distances between those in the field of graphics or web design, crystal ware, furniture, cars, hi-fi equipment, and kitchenware, to suggest a few to hint at the diversity. In some instances, attributes like energetic, logical, organic, dense, contrast, and several more, may apply.

The balancing of these factors may be decided by aesthetics and form at one end, and functions and technology at the other. Form equates with ways of expression or language, which may be influenced by other designs, nature, etc. Reducing these principles to practice resorts on several of the elements mentioned, such as style, rhythm, depth, and stance. Finally, delivery encompasses packaging and installation guidance, but also showrooms and support.

Taking an even more abstract view, a design or a product, service, etc., may also be seen as being built in a number of layers. One layer would be the physical one, its concrete construction, and another its performance, what it does. Then, there is the form or the aesthetics, how it looks, sounds, and so on. There is, moreover, the overarching mental layer, the meaning, and a behavioral one, how what is designed interacts and feels. 'Meaning' may be of two kinds; first, what belongs to the sphere of the culturally established, tradition, and perceptually given — round figures less aggressive than edgy ones — and, second, what is shaped by ideologies and religious beliefs.

If those were some of the minutiae of design, building blocks to design might furthermore be seen as not only the software increasingly associated with product or technological systems, but services too and, yes, experiences — and, by extension, processes, policies, organizations, communications, activities, events, attention, and emotions. These are factors dealt with elsewhere in the book. Let us just recall that a HD motorbike would be seen as an incomplete design without its — patented — roar.

A complex product like a car or an aircraft consists of components, modules, and subsystems at several levels. For a new model, some of these are to be new and unique to this new model or, at least, platform. The entire design process can be seen as an intricate web where the individual design problems are illustrated with the waterfall model: when the entire design is complete, this may be the time and stage where signals for change emerge to be subject to feedback to the very earliest development steps. A useful model and method for such projects is the 'Design Structure Matrix,' developed by Steven Eppinger and others. In this context, we need to outline it only superficially. The design process as envisaged is structured in discrete steps, each corresponding to a subsystem in the entire product or system. The interdependencies, also indicating something about the interfaces and 'buses' involved, are marked up in a quadratic matrix with the different steps both horizontally and vertically, the interdependencies as dots in the matrix. Let us just mention two important features of this description. It is now possible to locate clusters of subsystems that are closely interdependent but not much connected outside the cluster. With these links highlighted, it may make much sense to rearrange the project somewhat chronologically, and to subdivide development tasks. Another feature is that one can see where there is feedback from late development instances to early ones, and see to it that these loops get closed as early as possible.

Product structure may also be described with a product grammar.[2] This can be derived through the systematic parsing of product structure, analogous to the parsing of language that results in language grammar.

Product and product family expression may be designed within the frame of what is called a design language, a signature consisting of form elements, carrying aesthetic, sign, and symbolic functions. We will provide but one example: the often-cited Snow White, developed by frog design for Apple Computer, in use 1984–1990.[3] It features vertical and horizontal stripes serving as decoration and sometimes ventilation, also creating the impression that the computer enclosure is smaller than it is. First used in the Apple IIc, peripherals and accessories in the same style followed. Initially creamy off-white, it was later toned warm grey. The name is not for any color but a project code name from the seven dwarfs fairy tale. It is significant that frog did not try to define the language verbally; this is too hard. The stripes resonate with characteristics of the Macintosh user interface; they contribute a sense of precision, measuring 2×2 mm and placed 10 mm apart; they can be extended throughout the plastic to allow for ventilation. They stop 30 mm before the edges, ending in a perpendicular line reinforcing the illusion of reducing the object's length. To achieve good precision, designer Hartmut Esslinger argued for choosing a more complex production process than any computer company had tried before; the more costly tooling actually turned out to be offset by reduced materials consumption. A further attraction was that clone makers could not copy this process.

There are, of course, myriads of tradeoffs in design work, and we have encountered but a few. One way of enforcing creativity, impacting heavily on design, is to require a drastic reduction in some important parameter, say product price, development time, or the number of components utilized in a product. One example was a computer printer where one such design goal was to reduce the number of parts from about 150 to 60 or less. Clearly, this had repercussions on design. Various subsystems may be integrated into larger components; a few much more powerful components may substitute a number of simple ones.

Each such choice involves decisions along the standardization-customization axle; each such choice comes with a cost tradeoff; and the choice impacts production and logistics, after sales service and support. These are hard choices and with everything we have argued about the benefits of standardization and open interfaces, there must be something original to attract buyers. That, however, may be transient, and the competitive advantage is, rather, a facility for rapid re-design, for always besting the catchers-up. Another success recipe is to spot not one but several, perhaps a handful — according to one statistics-rich study — distinct advantages: their cycles of transience will not be synchronized. In other words, an entire system, where the various auxiliaries (often services, methods of selling, financing solutions, etc.) surrounding the product offering make that product much more than a 'simple product'. Or else we have succeeded in designing a true classic, which stays unaltered in the market just because it is *so* enjoyable, *so* much of delight to the customer.

Choosing among available components for inclusion in a modularized design is no simple task: it is always the performance of the entire system that must be in focus. As we saw previously, it has transpired that products that sport an outstanding performance on one single prominent feature, where top results are hard to achieve and, therefore, are much in the spotlight, actually loses in the market to products that feature an optimal performance along all relevant dimensions.

That reduction in parts from 150 to 60 obviously has to do with the complexity versus simplicity dualism. There is a plethora of measures for complexity, many of them associated with mathematics and algorithms. It might be an idea to develop a metric for the complexity of *your* products and to use it when judging the merits of new design suggestions. In many situations, such a metric would be a mirage as long as we expect numerical values, but the considerations leading to qualitative judgments and comparisons may well prove to be highly valuable.

The number of components and sub-systems, like in the printer example, is one metric. Plotting the product structure as a hierarchy may provide another helpful measure.

Thus, this chapter ends with a suggestion for measuring the effects of modularization, assisting also in getting more to grips with the various tradeoffs involved. A somewhat related question is whether there is to be a family of products, and what modules, design components, or themes should be recurrent or interchangeable.

Some of these decisions are related to your particular situation or task: to the extent to which you rely upon product platforms, are initiating and, in effect, choosing the design for new innovative architectures — or extending a product family, updating technology, modules, or sub-systems more incrementally.

In 2002, the Swedish producer of housing for electrical network stations KL Industri AB was in a crisis, profoundly loss making. Every house was tailored for the customer, costing design and production time as well as a massive handling of components. If the company was to survive, something radical was needed. The goal became to reduce production time, as well as the number of components needed, by half. Company employees, afraid of losing their jobs, had to be instructed about the value of design. The idea was that the exterior should permeate the feeling that contents were precious and dangerous but the little building should also give a nice feeling. Instead of tailoring just everything, modules are added to cater to specific customer needs. KL Industri moved from the brink of bankruptcy to comfortable profits, with revenue growing threefold, and the number of employees substantially increased.

Endnotes

---

[1] http://www.kk.org/thetechnium/archives/2009/06/triumph_of_the.php.

[2] Chin, Ryan C C, 2004. *Product Grammar: Constructing and Mapping Solution Spaces*. Master's thesis, MIT, Cambridge, MA; Chin quotes heavily from a series of papers of William J Mitchell.

[3] The paragraph on Snow White relies on Wikipedia.

Chapter 10

THE FUTURE IS — NOW

This chapter speculates about what is in store for design and designers today and in the future, and, consequently, for innovation, and design-related innovation. Some of the suggested trends and novelties may be regarded as wild cards, whereas others are well on their way to becoming established. As the saying goes: the future is already here, it is only unevenly distributed. Some of the tools, systems, and insights introduced in previous chapters or in this book's forerunner, *Design-Inspired Innovation*, certainly merit further consideration because of their potential for being transformed or augmented.

Fundamental changes in our very appreciation, views, and understanding of design would have drastic repercussions. Can we foresee the transformation of the already doubtful — in singular — design process into one or several *families* of processes? What about notation systems for design, for a number of the variables relevant to describe and define design? Will systems be developed to allow end users to design a larger part of their objects — and services? — in ways somewhat similar to that in which IKEA has made the customer undertake the final assembly of furniture? We design for sustainability, with a view to what happens during the product's entire lifetime, and environmental effects also after that, but will we see, possibly, a life cycle where the product is re-designed regularly throughout its life, and, if so, by whom?

Some of this is already happening, with rapid prototyping and Internet availability. Ponoko is a company but also a business model that offers direct contact between consumer and designer. It serves as an online location for making, sharing, selling, and buying products and product designs, for providing technology and materials. Ponoko claims that the user-designer will be able to produce just about anything relying upon Ponoko — at a mouse-click. The site hosts open source as well as commercial designs. The core vision is a *trade* in product designs — like the trade, paid or free, in music (iTunes), photos (Flickr), movies (YouTube), and software apps (iPhone and others) currently popular. Tens of thousands of user-generated product designs are hosted, ready to be customized and made into tangible products with that mouse-click. In addition, Ponoko provides a digital pre-production and production system so that the product designs can be priced instantly online and made locally, close to the point of consumption. Manufacturing on demand reduces logistics — transportation but for that of the bits!, warehousing, and also wastage.

Zoybar is another example, an open-source hardware platform dedicated to musical instruments, letting anyone invent their own instruments. Zoybar sells kits through its online community, the raw materials for making instruments (Fig. 10.1). In 2009, there were dozens of inventions, from a t-shirt that doubles as a piano to a guitar equipped with at touch-screen interface — should we think of the ordinary strings as an interface?! — and a built-in special effects facility. Any independent developer-designer can hack, modify, and attach third party components, with Zoybar's open hardware offering a shortcut, the web community allowing for the sharing of ideas, models, and experiences: feedback.

James Dyson, the inventor of the cyclone vacuum cleaner, envisages a time when these principles will be more generally applied: there will be open design software, open source drawings, and an open source library of components, 'published' by designers. An on-line library might offer downloads of components into a design which is then translated into reality through rapid prototyping.

Open Source Hardware, or OSH, is defined by the practice of openly shared documentation that illustrates how to design and build electronic systems of one's own. In addition to document sharing, the

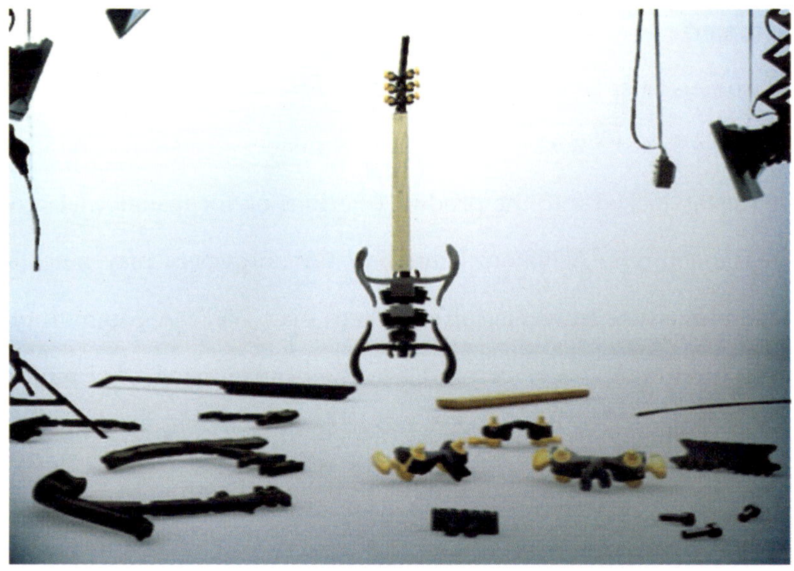

Figure 10.1.

OSH community is in the process of creating modular electronic building blocks that make it easier for lay enthusiasts to develop their own customized gadgets, like a GPS-equipped camera or an accelerometer. Dean Kamen's Segway has inspired Honda as well as hobby enthusiasts to create new designs with gyroscopes providing a key function, or several. The profound human quality that Kamen focused on was the human faculty for balancing.

The self-made gadgets constructed relying upon open source then return customized information flowing back to the community of enthusiasts. Consider Bug Labs, which provides a consumer-friendly electronic platform that can be customized into desired end functionality. This is a profit-making company, relying upon a mixed business model of free plus cost. Using additional modules like GPS, mp3 player, keyboard, and camera — at the time of writing, there are 80 such modules to pick from — snapped onto the platform, each user may customize the exact gadget as she wishes, and when it suits her. Consumers no longer have to stay locked into the product offerings of, for instance, electronics manufacturers. A Nikon, Pentax, or Canon camera may not allow for the immediate transmission of photos over Wi-Fi. A community of enthusiasts might "hack" or re-design the camera using OSH building blocks well enough to install Wi-Fi functionality in the camera. Question: what about warranties? Look out for firms adjusting to customers' demands! This hacked camera would then upload photos to the web, like a cell phone.

NKS — a "New Kind of Science"

After this introduction, highlighting some trailblazers for open source and sharing, we start with a seemingly exotic subject. In 2002, Stephen Wolfram of Mathematica fame and success (and later with the search, no, solution engine Wolfram|Alpha), published a gigantic tome of almost 1300 pages, somewhat grandiosely entitled "A New Kind of Science".[1] Wolfram has systematically reviewed different rules for generating generations of cellular automata, a couple of hundreds (256) for two-colored, two-dimensional ones, something which yields intriguing geometrical patterns, intriguing because they much resemble phenomena in nature. So Wolfram's bold contention is that this is another way by which nature can be understood and analyzed, based upon his suggested principle of *computational equivalence*.

There are regular conferences on the subject, and a fair number of other researchers are venturing to tread the same path. Many of the results generated can be followed on an on-line forum with an immense number of contributions. In addition to patterns in nature and culture — shells of crustaceans; Byzantine and Muslim mosaics, and abstract decorative patterns (Fig. 10.2) — attempts to apply the approach to other areas, like economics, are emerging. Three-dimensional objects can be formed by stacking two-dimensional successive cellular automata patterns on top of each other. So, what about design and innovation in the future? Already at an NKS conference in Boston, 2004, Rafal Kicinger and his colleagues at the George Mason University demonstrated early work on NKS and also evolutionary computing applied to structural

Figure 10.2. Tea cozy resembling a cellular automaton pattern.
The cellular automaton consists of a line of cells, each colored black or white. For every step to a new line of cells, there is a rule that determines the color of the cell in the next line from the color of the starting cell and its immediate left and right neighbors on that starting line.

design, whereas Brian Moulton, of Brown University, has applied NKS to the design of new materials.

Since each cell can be black or white (in a 2-color cellular automaton), this allows for two to the power of three (that is, eight) possible combinations along the top three cells. Because each of these combinations will cause a cell to be either black or white and there are eight possible upper color combinations, there must be two to the power of eight (256) possibilities in total.

Cellular automata seem utterly lifeless, dull, and lacking of interest at the outset. The wondrous thing, which caught Wolfram's

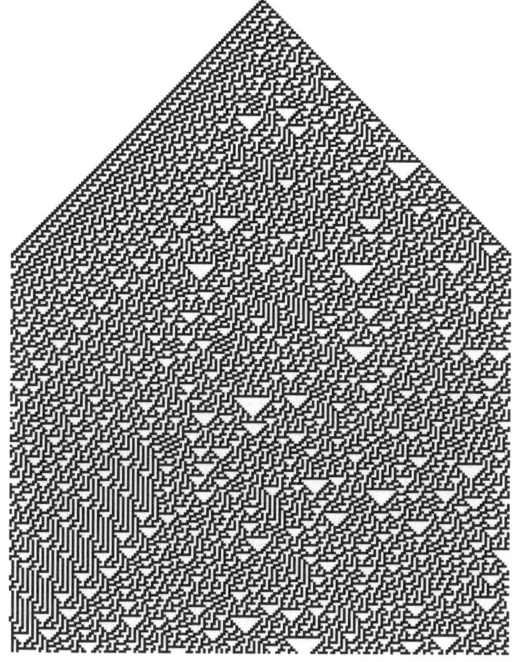

Figure 10.3.

fascination — and not just his — is the fact that simple rules may generate surprising, interesting, and, yes, beautiful results (Fig. 10.3). Regularity, beautiful patterns, eruptions, projectiles fired off — or, sometimes, dull repetitions or utter chaos. The basic characteristics, a few simple components although many of them and a few simple rules, also became the founding principles for a new field of mathematics, complexity science, with wide applications, the field of economics included.

So out of utter simplicity stems complexity: patterns, geometries, figures, beauty. That these should arise cannot be deduced from those simple rules and components, is not subject to reductionist analysis. The results are termed *emergent structures*, operating on a higher level than rules and components. Complexity mathematics has been employed to analyze an economics phenomenon such as the Silicon Valley, generating profound insights; more generally, in the field of increasing economic returns. Silicon Valley is certainly not designed, but is an emergent phenomenon, and so we may postulate that future designers will attempt to design for emergence to evolve.

The Internet

The World Wide Web seems to offer an immense arena for innovation. Currently, one craze is with social media that let us establish communities and share opinions and much else in new ways. We have met the suggestion that creating or, perhaps even better, having design users originate and design their own communities may be part of the larger scope of a design effort (Ponoko and Zoybar above; Nike+ and KOR among our case stories in Chapter 1).

In 2008, John B Rogers opened a Web site, Local Motors, where people are invited to contribute towards designing a car — inputting their dreams. They can also enter competitions for ideas for specific types of vehicles. The online community votes on designs, and winners are thus chosen by the audience. By mid-2009, there were 4,000 contributors to the site, providing drawings, comments, and voting. The total presence on YouTube, Facebook, and Twitter consolidates to 50 million people.

What more with future novelties after My Space, LinkedIn, Twitter, and Facebook? What about emergent phenomena in these spaces? More generally, Internet development is captured under construed but somewhat agreed upon headings such as Web 2.0, Web 3.0, etc., and it is important to try to see the essentials in each "new" generation, even if the transitions are not generational but evolutionary. What are the effects when brains can communicate directly with computers — something that is already possible but not yet easy and accessible in a broader way?

Consider a cultivator's tool — it exists — that Tweets the moisture status of the soil of a plant, reminding the owner to water it.

The 'Internet' of things

One trend promising huge change potential is the one associated with equipping objects with the capacity to speak for themselves, and to communicate with one another, on their own. That is what security tags and RFID markers already do, like in Coca-Cola's new vending machine, but the components and systems in the making will be physical entities that employ more sophisticated digital entities and sensors to make them call for some kind of service or to signal something important. These are signals that sometimes may be received and acted upon by other *things*, components, machines. The entire system and its associated functionality and service offerings will have to be designed, however. Some examples of this budding development include Pachube that, on its web site, claims:

> "If you have a device, building, environment or sensor (either physical or virtual) that is connected to the internet (wired, wireless or via SMS gateway) and you want to **store, share, graph and distribute its datastreams in realtime... ...**then register a feed!"

... and then, as to the output:

> "If you have a device, building, environment or actuator (either physical or virtual) that is connected to the internet, or if you have a website, and you want to **embed, monitor or connect to an existing environment...** ... then use a feed!"

A lot of current Pachube (Fig. 10.4) projects are using Arduino, an open-source electronics prototyping platform. Another example is the French company Violet that offers Mir:ror as their comprehensive tool for establishing the type of ubiquitous communication we are talking about. IPSO is a standard aspiring to become *the* standard for the Internet of things: the IPSO Alliance has announced "the start of the first Certification and Testing Compliance program for embedded IP and IP Smart Objects, such as those used in Smart Grid, Home Controls, Building Automation, Industrial Process Control, and Health Care."

Of course, one of our initial cases — Nike+ — might be seen as qualifying as one of these connected 'things', though a fairly simple one.

'Things' should read places for Sense Networks, which indexes and ranks locations in the world based on data for movements between these places at different times. These movement and location data are collected in real-time from devices with GPS and WiFi positioning technology, such as mobile phones, cameras, and cars. The resulting analysis gives a rich profile of locations in a city, to better understand visitors and their needs. The first application, Citysense for a city, is being installed.

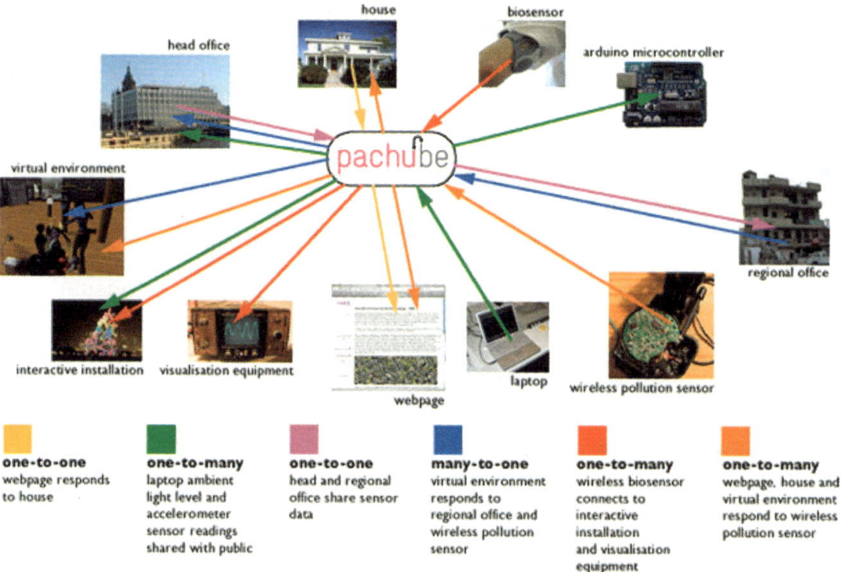

Figure 10.4.

Hewlett-Packard works on a more general effort, the Central Nervous System for the Earth (CeNSE), comprehending a trillion nanoscale sensors and actuators embedded in our environment and connected through a vast array of networks with computing systems, software, and services, exchanging information among analysis engines, storage, and users. The ambition is to impact human interaction with the globe in the same way that the Internet has affected personal and professional interaction.

Biomimicry

Velcro is the most popular example of biomimicry, a phenomenon which we have already met as one of the ways to stimulate original thinking — creativity. Finding analogs and parallels in nature may be greatly extended in the future. Some species exist in symbiotic relationship with others, others in parasitic relations. Likewise, symbiotic and parasitic business relationships already exist, and it might be feasible to model such relationships or systems; in the future, we may come to design for their co-evolution. This would require tools for visualizing such systems so we can suggest the future development of analogs to CAD and scenario-writing and feedback systems descriptions with a view to symbiosis.

When Eric Drexler[2] wrote about nanotechnology and "engines of creation," he postulated mechanisms that were self-assembling and replicating. The idea may meet fundamental obstacles but it points to the potential emergence of what we can call 'meta-design', designing a *mechanism* for performing design, possibly with feedback loops included.

To take Drexler's proposition to the field of life instead, it has been suggested that a future form of design may be that of designing genomes to make for new varieties of plants and of animals. This will surpass what classic plant and animal breeders' have been doing — by far. According to this proposition, new berries and trees, new mammals and fish would be created much as fashion, cell phones, cars or chairs are designed today. Dream or nightmare? And what about ethics? This would certainly mean a new vista for design. Again with new demands for tools allowing the designer to better foresee the results of interventions and of design acts. Such new or tweaked species may need a particular context, also designed, to thrive and be useful. Another possibility is designing relying upon protocells, not quite life, which can be 'directed' to grow in useful directions; there is currently a project where protocells are used to strengthen the some 1,200 year old timbers that constitute the foundations for Venice.

Our senses

Some of the wildest cards may be associated with our faculties for tasting, smelling, and sensing. Our reactions to odors are believed to be very powerful but largely unknown and partly unconscious; even the mechanism of smelling is not well understood. So a breakthrough here may be something to act upon. Feeling, our tactile sense, is not one but several — our haptic senses, where our outer skin is our largest sensory organ in a physical sense. We feel mechanical pressure, but also hot and cold, we feel pain and vibrations, and we have a sense of balance versus its opposite.

Thinking traps

When Daniel Kahneman received the Economics Prize in the memory of Alfred Nobel, it was to honor his work, often with Amos Tversky, in behavioral economics. Kahneman and Tversky had — and they are not alone — pinpointed a number of behavioral traits that makes the idea of 'rational man' highly doubtful. This is not the place to dwell on the entirety of this body of knowledge, only to give some examples of why doing so is important and why it probably may be incorporated in future tools and practices of designers.

One example is anchoring. Ask someone in a group to suggest a number between one and 100. Then let someone else guess the number of African states in the United Nations. If the number was low, the number of states suggested will be low while a larger number yields a larger number of nations. Another example is endowment. Give each, in a group of MIT students, a mug emblazoned with the school's logotype, and then have someone ask to buy the mug. Compared to what students who have not received any mugs would pay for it, the price requested for making the endowed student part with *her* mug seems ridiculou sly high.

When Dyson challenged conventional wisdom in vacuum cleaner motors (see box), he was also avoiding yet another thinking trap: to be satisfied with something working 'reasonably well'. A Swedish company had the majority of the world market for a specific type of heavy industry machinery because of technological superiority; on e day, a competitor had completely turned the tables. The Swedes' ire was that the new breakthrough had nothing 'new' to it; it was just a clever application of longtime existing knowledge, not really contemplated before. The market leaders had stopped pondering better ideas, because they had thought that what they had was good enough for market dominance.

> James Dyson of cyclone vacuum cleaner fame was dissatisfied with the motors available for his vacuum cleaners. Thus, he established new requirements, a design dream if you wish, including quantum leaps in life and reduced power needs in his own lab.

The resulting new design featured a 33 percent increase in sucking power followed by an even better engine a quarter of the size of its predecessor, implying, e.g., a doubling of the battery life of the handheld vacuum cleaner (Fig. 10.5). In a sense, the future is now, when established rules of thumb and given solutions may, and should be, questioned.

Rules of thumb? This is the principle of master inventor and electricity systems professor Mats Leijon: always discard rules of thumb, because they reflect practical stumble-stones and a state of the art that constitutes history. Instead, investigate what the fundamental limits are, those established by laws of nature, be they Newton's, Maxwell's, or Reynold's laws.

Some thinking traps have to do with our utter inability to handle dependent probabilities. To offer examples would be to derail the discussion because experience shows such examples to challenge intuition, which is an indication that intuition sometimes misleads gravely. Gerd Gigerenzer[3] and others[4] are developing tools, 'simple heuristics that

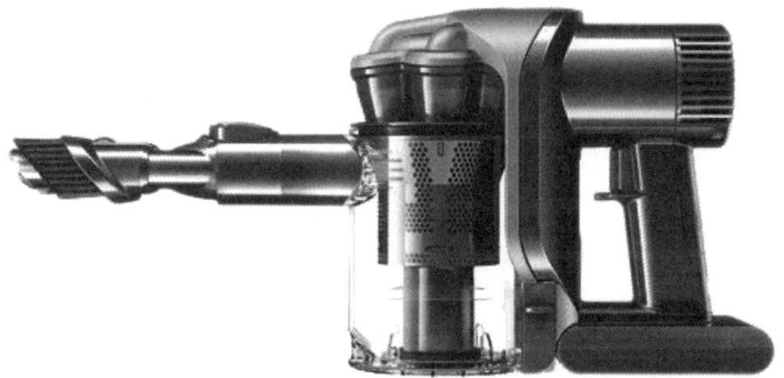

Figure 10.5.   The Dyson engine in a handheld vacuum cleaner

make us smart,' that offer ways to best those dependent probabilities. The challenge for future designers will be to translate such heuristics into *designs* that make users smart, and, perhaps, even to develop new heuristics, the design way. Langer[5] quotes Gigerenzer's demonstration of how probabilities may be expressed starkly differently: the *relative risk* of using mammography means reducing the risk of dying by 25 percent since only three women instead of four out of one thousand would die. The *absolute* risk reduction in this case is one in one thousand. Or: to *save one life*, obviously one thousand women have to go through the procedure. The result of saving one life, instead included in the *life span calculation* for those one thousand women, would be all of twelve days.

A particular type of probability conundrum is the one associated with what should be regarded as 'typical' or 'normal' in a broad, general sense. Perhaps we should rather discard such notions altogether and treat people and eponymous 'end users' just as individuals. But the idea of something 'average' is bound to re-appear — often enough, unfortunately. One way of questioning this idea is to compare the perceived average with the perceived mean value. Another way would be to test whether Chris Anderson's 'the long tail' argument holds. But the most fundamental issue would be to raise Taleb's Extremistan versus Mediocristan argument; the normal distribution is very far from *normal*, and the median, if it exists at all, is hard or impossible to come by. Again, how may, and how will, future design come to the rescue?

Psychologist Ellen J Langer has performed a number of studies underlining the importance of 'mindful thinking' and also demonstrating how such thinking is often undermined unwittingly. Once more, this is not the place for a recap of a book, of Langer's book, *'Mindful Thinking'*,[6] so, again, an example is called for. In the US and Western Europe, we are well aware that memory loss befalls us when we get older — and testing bears this out. Langer tested a very special North American culture — that of the deaf. Here, the elderly are revered and the idea of memory loss with age does not exist; nor does it exist when tested for. The next culture is not so small: the Chinese. No pre-conceived conceptions of memory loss with aging, and none are recorded either.

Self-fulfilling prophesies are just that, and mighty they are. Which will be the next one to be revealed, and how to design to reveal or avoid such prophesies? Langer provides a number of strategies and ways of how to phrase problems that help avoid pitfalls, and to get into the 'mindful' mode instead.

Innate behavior

We have underlined the fact that words tend to overshadow images and visual memories; we have told of recent experiments demonstrating that different languages induce different modes of thinking. None of these discoveries were foreseeable through a discovery matrix or morphological map. So we will just suggest that we must be on the look-out for more such intriguing exploits, relevant to the understanding of how people act, and thus to how we design.

Will we be able to profit from emotion detectors, reliable ones, when trying and testing out designs? Developments are under way, following different avenues. Moods and emotions are reflected in face expressions that may be registered by cameras, and these recordings then analyzed by computers. Fleeting expressions, like a lip briefly curved, may be highly revealing. The software for this analysis is still wanting but improves steadily. Voice is another betrayer of emotions and emotional changes. Psychologist Paul Ekman has identified seven universal emotions: sadness, happiness, fear, anger, disgust, contempt, and surprise.

Emotion detectors: during the fall of 2009, Dutch electronics firm Philips and the ABN Amro financial company introduced the prototype of a bracelet that, by measuring the electricity going through the skin, can warn the one wearing it of emotions likely to carry her away (the same technology as used in a lie detector). Gestures also tell of emotions and are depending upon culture. More speculatively, and most probably more into the future, would be smell: we are known to generate minute amounts of chemical substances, for example, when getting into a panic or simply being afraid. Animals perceive this!

Network mathematics

It is a field that has generated a lot of interest and new insights lately. It may apply almost wherever there are networks: to the Internet, telecommunications in general, and logistics, such as roads and air traffic routes, genes, neural nets in our brains, food chains, relations between people such as actors — and, of course, within communities engaged in innovation and design. Such systems quickly turn non-linear in a mathematical sense, which has the consequence that they do not lend themselves to reductionism; they feature emergent properties.

So those persons designing logistics support systems and value chains may be assisted in making these robust and self-repairing; the genetic engineer may spot critical genes for better health or higher yields; the health systems designer may map the contact networks to locate key nodes for contagion, to change that network's design to halt epidemics before they have become such; the urban designer may resolve or prevent traffic jams; food and value chains may be redesigned to rescue threatened ecological systems; social networks coalescing into communities may receive critical interventions; infrastructure for developing countries or just in general may be better optimized, value chains perfected; networks at large designed for resilience and robustness.

There is, for example, a great difference between egalitarian networks, where each node carries about the same 'traffic,' and aristocratic ones, which are more hierarchical, with central nodes that are much more busy than the others. This is one important distinction; another pertinent lesson is that *weak ties* turn out to be the strong ones, the important ones, in the sense that it is not the regular network of a firm or a person that comes to the rescue in an unusual situation, but the rare, more peripheral and, therefore, regarded as weak ones, thus the gateways to the outer, more exotic world, the foreign networks. A third observation is about networks dynamics: when networks are sparse, they behave in one way, loosely connected, but when they get denser, they all of a sudden reach a transition point where they coalesce. It is like what in physics is called a phase transition, such as when ice crystallizes instantaneously in water still not frozen but under the freezing-point, or when whipped cream, when whipped too much, solidifies. Emergence, if you like.

Super crunchers

Ian Ayres commences his book, *'Super Crunchers'*, with the tale of Orley Ashenfelter who, based upon statistics, devised a formula for the quality of Bordeaux wine.[7] It rests on just three factors: winter rainfall, average growing season temperature, and harvest temperature. Anathema to wine aficionados, this formula bests them and their noses and palates all. The point here is statistics: what you may discern from available piles of numbers, to which computer memory capacity and computation power allow access and crunching. Closer to design, there is a formula, based upon New York Times Bestseller Lists, for testing various suggestions for book titles; Ayres' 'Super Crunchers' fared reasonably well as a title, based upon this system, but to arrive at this he had first used another potent super crunching mechanism, Google and Google AdWords. Anyone searching for 'number crunching' or 'data mining' was shown 'Super Crunchers' or 'The End of Intuition.' 'Super Crunchers' won, 63 percent to 37, and the duel outcome was settled. Opportunities for designers already exist and are bound to multiply, with lots of new Internet facilities and more and more ingenuous ways to acquire and crunch data. Huffington Post uses the same technique (called A/B testing) for some of its headlines; it makes it possible to design differently for different audiences, such as the West coast versus the East coast of the US; Amazon and Google belong to the pioneers with this technique.

A supercrunching example is the first Netflix Prize awarded for an algorithm producing improved movie recommendations based upon previous movie pickers' preferences. The data set and the evaluation metric were the two important starting points, the 'dating' between different teams — 40,000 from 186 countries entered the competition — finally making for the necessary combining into super-teams of people who did not know each other at the outset. The winners were four teams from four countries combined, unknown to each other before the competition, each contributing something unique, one of them describing elements of movies in a fresh way — analyzing design, as it were. Will it be possible to go from preferences of what does exist to the design of what does not?

Microtrends

This is another take on the availability of numbers, and number-crunching. Mark Penn is heading a market research company, Penn, Schoen & Berland, providing him with empirical support for the many niche trends spotted in his book, *Microtrends*.[8] There are three arguments following his book introduction. First, that important trends are small before becoming dominant, then, that there may exist important niche markets left undiscovered and drowned out by dominant trends ("America is moving in hundreds of small directions"), and, third, that there are emerging counterintuitive trends that are powerful enough but overshadowed by events and forces stealing the spotlight. As an example, he contends that teen crime tends to conceal all the young people who are actually succeeding as never before. Focus has been on poverty as the cause for terrorism, while it is well-to-do, educated terrorists who have been behind many of the attacks. An important part of mobilizing interest in a newly discovered niche group is to give it, to design, a proper name, like Penn's prototypical 'Soccer Moms', a persona of sorts, though an exceedingly brief one.

Memetics

The meme concept is in itself a meme. Richard Dawkins[9] coined the concept in his 1976 book, *The Selfish Gene*, as culture's equivalent to the gene. As examples, he suggested 'tunes, ideas, catch-phrases, clothes fashions, ways of making pots or of building arches.' Other attempts have been made to come up with a word for culture's gene equivalent, such as culturgene, but it is meme that has taken hold. More than Dawkins himself, philosophers like Daniel C Dennett and Susan Blackmore have contributed to the putative field of memetics, the science of memes. Blackmore has even suggested that there exists a third replicator that is fully technological in nature, frequently flourishing on, e.g., the Internet; she first suggested to call it 'theme' but later asked for better suggestions.

Like the gene, the meme is a replicator. To be a successful one, it has to display fecundity, fidelity, and longevity. A few distinctions might be useful. First, even a gene is not well defined in the sense that there is a sharp cut-off between one gene and the next part of the DNA spiral, and the meme is not always so clearly cut-off either. For the larger systems complex called the human *genome* in biology, the memetic counterpart has been variously named, among others, 'memome' and 'memeplex'. Second, if we think of the arrangement of a string loop called 'cat's cradle,' no child's cradle is exactly the same as the other's: the fingers and hands are not identical, nor are the strings' lengths. Still, cat's cradle is a meme and the important thing is that the children follow the same recipe; it is a case of 'copying-the-instructions' rather than 'copying-the-product', and the former makes for much better odds for fidelity. There exist, of course, descriptions on how to 'unleash the idea virus' and how to manage and operate through concepts and meaning.

Genes change only infrequently and the results show up on long time scales; how long is disputed. By contrast, human culture, socially constructed, may develop or change much more rapidly. A meme may display fecundity and the required fidelity, but what could count as longevity may depend on which part of culture is involved. Tunes may become hits without demonstrating the same longevity as the four famous notes of Beethoven's Fifth Symphony. Media play a role, to which it is hard to find an exact equivalent in Darwinian nature. The fact that genes are physical may be seen as at the core of the sharp criticism that has befallen the idea of memes. Do they really exist in any meaningful sense or do they constitute a misleading mirage? If they proliferate, do they not do so rather in a Lamarckian learning mode than a Darwinian one? The point here is to point to a future when these questions have been resolved. If ever memetics becomes a trusted science, it might imply that designers would have reliable laws to depend upon for judging the merits of a novelty, and the incentive to further the development of more such laws. It is a big 'IF'.

Neural networks

A particular type of number crunching — super crunching it has to be — is the one undertaken by neural networks. These are computer systems that emulate the neural nets of the human brain, with a capacity, like the brain, to learn from experience. But are there enough of learning experiences out there for them to be trained to provide any genuine contributions? This question, thus, asks whether there are sufficient amounts of data to describe experience in minute detail. Yes: Epagogix is an American company prejudging the success capacity of movies, breaking the iron law of moviemaking that nobody can foresee what will be successful. The system is not perfect but scores well already when the movie is just a script, no stars and no director contracted or penciled in. The next putative step would be for the firm to actively participate in making scripts, to use those neural networks interactively. Thriller plots, movie storylines — what is so different from many a design?

Interactive genetic algorithms

As in other fields of problem-solving, Darwinian methods have been suggested and are employed for design tasks, another field of promise for the future. The formal name is Evolutionary Computation (EC), a search technique relying upon computational models of processes of evolution and selection. EC utilizes a host of terms from biology applied to, e.g., the individuals within the populations considered. Concepts and mechanisms of Darwinian evolution and natural selection are encoded in evolutionary algorithms (EAs), and applied to solving problems. Several varieties of EAs have been proposed and studied. They share the same underlying assumptions but differ in breeding strategy and the representation on which EAs operate. There are four major types as well as hybrids between them; they are called Evolution Strategies (ES), Evolutionary Programming (EP), Genetic Algorithms (GA), and, the most recent one, Genetic Programming (GP). The latter has been employed in developing functioning computer programs.

Like in nature, EC is a search and optimization process where a population of solutions undergoes a process of gradual changes. The process proceeds based upon the fitness (a subjective or objective measure of perceived performance) of the individual solutions. A typical EA consists of the following steps, repeated until saturation or no significant change between results from one completed series of steps to the next:

1. Initializing the population.
2. Evaluation of all population members.
3. Selection of individual(s) in the population to act as parent(s).
4. Creation of new individuals by applying variation operators to copies of parent(s).
5. Evaluation of new individuals.
6. Replacing some or all of the individuals in the current population with new ones.

So, first, an initial population of individuals (solutions) has to be created, possibly randomly, but other techniques are possible, such as using previously known or arbitrarily assumed solutions. Next, each individual in this initial population is evaluated and given a fitness value. The variation operators in step 4 are most often *mutation* and *recombination*. Mutation acts on a single individual, applying some variation to one or several 'genes' in that individual's 'chromosome.' Recombination, though, operates on multiple individuals, combining parts of these to create new ones.

There are examples of Evolutionary Computing being applied to model real world problems apart from computer programs, such as for structures or architecture, floor planning, and also user interface as well as document design. One challenge is to establish criteria for fitness judgments, which may well be multi-dimensional. There are examples of objective criteria — in an architecture example, room sizes, dimensions, etc. — as well as of people ranking different features — appealing–unappealing, traditional–revolutionary, meets all requirements–does not, etc. — to arrive at a composite fitness score. This screening of the population is the stepping-stone for generating the next.

Evolutionary Computing may be said to feature:

- little need, if any at all, for a priori knowledge of the search environment;
- excellent search capabilities because of efficient sampling of the design search space;

- robustness over a broad range of problem classes;

- ability to handle high dimensionality;

- provision of multiple good solutions;

- ability to avoid local optima;

- ability to locate the region of the global optimum solution.

As every design problem comes with some constraints defined, it is important to have these taken into account in the evolutionary process. One of several ways, and the most common, is by introducing a penalty function. Likewise, designs that must meet multiple objectives have to be treated accordingly, the process tweaked. One of the vistas for future development would possibly be the development of tools and interfaces that conceal the intricacies of Evolutionary Computing, making it more like handling a Lego set. A further sophistication is the introduction of co-evolution where requirements and design solutions develop interactively, or co-evolution in the sense that a design evolves in synergy with its environment, eco-system, or context. To combine our potential futures again, let us note that Genetic Algorithms have already been applied in the field of memetics.

— 0 —

It was suggested that this chapter should be speculative. It may seem to be displaying the result of some shotgun approach, travelling widely, bringing up an almost surrealist collection of disparate items.

In fact, many of them might not turn out to be so disparate after all. Neural networks, genetic algorithms, and *A New Kind of Science* are interlinked, and the first two have been applied to memetics. Biology, economy, and ecology are amenable to the treatment of network mathematics, as is evidently the Internet — thus the 'network of things' is, too. Another strand, though not lacking of connections to memes, are the Kahneman-Tversky and Langer 'thinking traps,' and, by extension, future breakthroughs in the understanding of human senses. Super crunching and micro trends represent something else, new methods to understand and interpret data; but that holds for network mathematics too.

This treatise has been about design inspiring innovation, and innovation is about novel combinations that serve a purpose. Not a surrealist collection of disparate items, then, but conventionally disparate pieces that design may bring together to new striking and elegant, delightful and beautiful entities — surprising and meaningful.

Let us suggest a neologism: *metovation. Meto-* in *metovation* stems from the prefix meta- where, for easier pronunciation, *o* has been exchanged for *a. Meta* is of Greek language origin and stands for 'the next higher level of abstraction.' Thus, meta-art is art dealing with art — metaphysics would be that kind of physics that organizes the more concrete physics, only that the term 'metaphysics' has acquired a somewhat special meaning.

Therefore, *metovation* equates with novelties, with innovations in the very ways of producing the 'ordinary' innovations, and thus *new conditions* for innovation. One obvious example is *open innovation,* something heralded in recent years. A Google search for *meta-innovation* results in the suggestion of 'systemic innovation accelerating innovation' — so a condition again.

Conditions for innovation: that would involve the very fundamental settings for business, industry, and entrepreneurship too. As an example, idealism and volunteering have always existed but the developments in *open source* and *freeware* must be regarded as contributing something fundamentally novel — when something is transformed from marginal to mainstream, it heralds a new dimension; quantity mutates into *quality*. What, then, may be the reasons for working for free, to generate resources for others to utilize and exploit? One human motivator may be conscience and generosity, another a realization that there is power in reciprocity and thus a plus-sum game — as I do, others will follow, which is to my gain. Yet another component may be reputation and track record, which may be translated into commissions and offers of employment. And there are other business twists, such as those who 'sell' *freeware*, that is, offer it for free but charge heavily for what might be called a user's guide.

It is hard not to see the Internet as a *metovation*. It may be claimed to be an innovation, an infrastructure, as well as marketplace. These alternate denominations capture different aspects of the Internet, and the very fact that several are needed makes it a *metovation*. The Internet has, for example, spawned Second Life where virtual creations (like avatars) may be exchanged for Linden dollars™, a currency that may also be exchanged for real money, real dollars. Linden Labs, the company behind Second Life, has created its innovation on the basis of the *metovation* — the Internet.

How would an entrepreneur, an inventor, or the managers of companies regard and treat *metovations*? As these examples highlight, *metovation* is most often something too generic to benefit just a particular individual or corporation — *metovation* offers new opportunities, and here it is up to entrepreneurs and inventors to latch on. 'Open innovation' may open up new channels for otherwise homeless ideas, while its close relative, innovation, from pioneer users may cause upheavals in competition space. Likewise for *freeware*. But new vistas by way of new business models open up, as the example of revenue-generating user's guides for free software or as Linden Labs currency demonstrate; the Internet and cyberspace positively brim with innovation. The Internet, Google, and many more also exemplify how sometimes the revenue model evolves or is invented only after the launch of an idea — and sometimes the expectation or hope that such a model will evolve turns out to be wishful thinking. Topsy-turvy business models is, of course, nothing entirely novel; a shaver for free to sell razorblades is a classic case, Xerox' nice, reasonable way of charging per copy is another example of a profitable prescription. So how would you regard β *design-inspired innovation* — a *metovation* to boot? The future has only just begun, tomorrow never comes...

Endnotes

[1]Wolfram, Stephen, 2002. *A New Kind of Science*. Wolfram Media, Champaign, IL.
[2]Drexler, K Eric, 1986. *Engines of Creation*. Anchor Books, NY, NY.
[3]Gigerenzer, Gerd, 2002. *Reckoning With Risk*. Penguin Books, London.

[4]Gigerenzer, Gerd et al., 1999. *Simple Heuristics That Make Us Smart*. Oxford University Press, NY, NY.
[5]Langer, Ellen J, 2009. *Counterclockwise*. Ballantine Books, NY, NY.
[6]Langer, Ellen J, 1997. *The Power of Mindful Thinking*. Da Capo Books, Reading, MA.
[7]Ayres, Ian, 2007. *Super Crunchers*. John Murray, London.
[8]Penn, Mark J & E Kinney Zalesne, 2007. *Microtrends*. Allen Lane, London.
[9]Dawkins, Richard, 1976. *The Selfish Gene*. Oxford University Press, Oxford.

Chapter 11

AFTERWORD: DESIGNING THE INNOVATION-INSPIRED MANAGER

As this book intends to be the reader's, I hit upon the idea of trying and testing whether a prototype reader would be inspired by it. That reader should not be deeply into design (not yet?!) and not an innovation scholar as such; rather, someone aware of innovation's importance but more geared at management in general. Mulling who might fit the bill, I came up with a great choice: Henrik Blomgren, PhD, dean and head of the department of industrial economics and management at the Royal Institute of Technology in Stockholm (my new/old affiliation). He is a person who has grown up in a small business family, a manufacturing concern he still takes a hand in at times. He, thus, combines a business with an academic perspective, spiced with experience from management consulting and a couple of years with the secretariat of the Royal Swedish Academy of Engineering Sciences, best described as a think tank, much concerned with innovation policy and, by extension, innovation management. As dean and head of department, teaching and learning are matters of major concern to Henrik's daily tasks. He agreed to serve as the test pilot reader, for which I am deeply grateful.

Setting the stage

To quote from the Oxford Handbook of Innovation: Innovation is *not* a new phenomenon.[1] And, thus: the need for innovation is not a novel topic of interest either.

During the years, at least since 1942, when the well-known economist and innovation theorist, Joseph Schumpeter, coined the term "creative destruction," academic interest in industrial renewal has, repeatedly, claimed that innovation is a necessity.[2] So companies trying to survive in the marketplace — thus all of them! — have incessantly been striving for innovation.

Still, innovation is a puzzling topic, and particularly challenging is the question on how to *manage* it: the task of innovation management. By definition, an innovation stands for something novel; therefore, no management precedents or prescriptions offer self-evident, proven advice.

New ideas, new concepts, new products, new business models, new services, new organizational settings, new trademarks, new designs, new styles, new colors, new flavors, new companies, even entire new industries; of course, they are necessary, unavoidable even — but still, how to achieve such renewal?

Today, it is often suggested that cycle times for change are shortening compared to, let us say, a hundred years ago. What is measurable is the abundant proliferation of new products and options on offer. Therefore, it is also often claimed that the current need for innovation is even greater than before. It is most probably no coincidence that today's discussion about the need for innovation sometimes creates feelings so intense and intimidating that we may experience ourselves as on the verge of paralysis, possibly even being scared by the discrepancy between a seeming lack of innovation, and all this discussion about its importance.

The notion that the speed of change in industry has accelerated merits being questioned, however. Professor and historian of technology, Nathan Rosenberg, has, for instance, demonstrated that mankind always succumbs to a feeling of living at a time of accelerating change, of exponential growth curves[3] ("chronocentrism").

The conclusion must be that we have nothing to fear from innovation. History tells us that it constitutes part and parcel of our cultural evolution; innovations have always been, and still are, natural components of our daily life. We have managed them before, and we have always succeeded in handling them.

This does not change but, rather, underscores the fact that innovations are necessary, now and in future. And since they are needed, since they are necessary, since they are unavoidable, there is an obvious need for their management.

Historian Peter Burke claims that change is a matter of how we *perceive* change.[4] Our perceptions of change are created in relationship to our perceptions of stability — and stability is perceived in relationship to how we perceive change. Therefore, the difference between change and 'steady state' is just a matter of perception. Perception is everything — a classic enough statement.

This also means that if we perceive a problem, there *is* a problem. If we do not perceive any, there is none.

So, since we are still feeling a need for the development of our knowledge on how to manage innovation, *it really remains* a topic of major concern.

Creating models for innovation management

On a corporate level, innovation and industrial change can be perceived (modeled) as driven either by what might be called external or internal factors — though, in reality, perhaps a combination of them — and may also be seen as, in a way, *managed* by these.

An external factor is something we, as corporate managers, do not have influence over, and, by that token, such a factor tends to exert a force on us: it can be said to manage us instead. It is something beyond the control of the company. Technology and politics constitute examples of such factors.

An internal factor is one taken care of within a corporation, perhaps by the entrepreneur him-/herself. Engineers, designers, and marketing people creating new kinds of solutions offer prime examples of internal factors.

Understanding technology, knowing the markets — basics for the innovation manager

As mentioned, technology is considered one important external force to innovation; though to some extent, it may be transformed into an internal one when adapted and harnessed for internal purposes. Or, like the rare case of the transistor for Bell Labs, in effect, originating internally for that company while becoming an external force for all other companies. Within the field of management, technology has long been considered as one of the most important external driving forces to innovation. The steam engine, the automobile, the airplane, the computer and the personal computer, the cell phone, and the Internet provide illuminating examples.

So far, engineers have constantly displayed faculties for managing innovation; it is not much of a stretch to presume that the future offers more of such a display. Thus, we may conclude that technological knowledge is a basic prerequisite for innovation management — even though seldom, if ever, sufficient for exerting "management control."

Marketing people, in particular, often describe the company Wal-Mart as a venture capitalizing on a need for cheap consumer goods in small, less developed — peripheral — towns and cities in the US; and during the last several decades, such needs have been growing. Wal-Mart is a profoundly interesting marketing phenomenon and today's books on marketing are full of stories on Wal-Mart.[5] Well, then, what was the marketing phenomenon behind it all – was it not Sam Walton the founder himself? That is a picture easily emanating from his autobiography, though the book is humble enough despite being authored by someone achieving such tremendous business success.[6]

Engineers often miss paying attention to (or discount at their peril) the marketing aspects of innovation.

So be it, but on the other hand, we might ponder where Wal-Mart would have been without the technology needed for, e.g., inventories, control of supply chains and logistics, and the engineers behind these phenomena? Would the company have achieved success without well-structured, sometimes even ingenuous, processes and huge amounts of technology internally? Most likely not: several have been held out as innovative and trailblazing — the plural important enough.

So the flip side is that marketing people often miss noticing the engineering aspects of innovation.

The former head of McKinsey Global Institute, Bill Lewis, has shown that Wal-Mart does not only passively exploit technology, the company is also a very prominent early adopter and implementer of *new* technology. Wal-Mart is important as a customer of technology — and to such a substantial extent that the company has even affected the GDP of the entire US.[7]

This goes to demonstrate that technology not only might impact markets, markets might also impact technology. Thus, we may conclude that market knowledge is a further prerequisite for innovation management — even though seldom, if ever, sufficient for exerting "management control."

Likewise, Lewis claims that companies like Dell, Intel, and Microsoft would have been *nothing* today without their major customers, who emerged and grew particularly during the 1970s — specifically Wal-Mart which expanded rapidly at that time. A picture of young heroic entrepreneurial engineers taking the initiative, however, still heavily influences received wisdom on how these ICT companies developed. Thus, we tend to forget the important role of forward-looking, demanding customers in making it all happen.

While broaching themes of technology and/or marketing, do not forget what might be behind it, what might *underlie* it — what might be found underneath.

Balancing technology and markets — a typical management task

In this context, how should we try to understand the emergence, existence, and development of companies like Skype, Amazon, Google, YouTube, Twitter, Facebook, Second Life, LinkedIn, Englishtown, and eBay — just to mention a few current phenomena created in the last decade or so? Are they instances of engineering or marketing? Or maybe they are cases of "design" — whatever we denote by that term?

Sometimes, design becomes synonymous with and restricted to "form," but a more preferable definition would be a far broader one — like "visualization of the possible."[8] So let us establish it as something at the crossroads of engineering and marketing — an idea that will be applied from now onwards.

A company having proven to manage this balance exceedingly well at that crossroads during the last decade is Apple Computer (since recently: only Apple) through constant renewal of interesting product lines, the retirement of older but not yet 'tired' ones, with services and technologies yielding huge customer appreciation (voted the world's most admired company in early 2010). It makes us curious about how they do it. Of course, management does play an important role when searching for the answer.

Assuming management would remember places of underneath

When evoking important external forces and drivers for change, the importance of basic natural science, mathematics, chemistry, physics, and biology, as well as applied science, biotechnology and nanotechnology, etc., also ought to be counted in.

Developing an understanding of the algorithms that are at the core of Google search is, for example, far more important than merely observing the fact that the company thrives on the Internet. In the book, "The Google Way," as in several others describing the company, the value of the founders' basic background in the fields of mathematics and computer science stands out very clearly.[9]

How could it be that Wal-Mart for decades could expand so easily and unencumbered in small, unknown towns, quite a few the size of some 5,000 inhabitants? How could it be that nobody else took an interest in that customer segment? Were all other retail companies fully satisfied with generating loads of money in the big cities? How is it that so few retailers saw the US demographic shift to suburbia?

Demography, sociology, psychology, history, art and design, and cultural understanding are all useful when trying to understand change, and to translate that understanding into a thriving business.

When Steve Jobs delivered his Commencement Address at Stanford in 2005, he suggested that an obvious connection existed between ideas in the development of Apple Computer and his studies of calligraphy as a young (drop out–drop in) student.[10]

Design, understood as form, is most definitely a contemporary force to innovation.

Having claimed that much, it is also reasonable to observe that services, new principles for organization, new types of business models, etc., may be just as interesting as physical products. We are no longer, at least not only, into the manufacturing of physical industrial goods in the 'industrialized' world; certainly not when it comes to generating value.[11] So innovations in those other categories are of importance too — sometimes tethered to physical goods. In that context, contemplating the contents of books like *"The Service Innovation: Organizational Responses to Technological Opportunities and Market Imperatives"* opens the mind for new vistas for creative ideas.[12]

Wal-Mart started off as a franchisee — franchising being an interesting organizational innovation in itself. IKEA is just as much (not to say more) of a business system as it may claim to consist of technological innovation. Ryanair is an entire concept heavily dependent on technology. Trying to understand Google from a business model point of view is really extremely interesting. How do they succeed in generating revenue and how is it that their business model so far has not seen any full-scale copycats? Design, interpreted as form, is definitely embedded in these companies.

If the singular focus dwells on bringing increased value to the customer (or potential customers), innovation could mean so many different things — *far* more than just technology and/or markets. And, of course, that goes for the management of innovation too.

Coping with turbulence

Two more recently highlighted, although long-existing, external drivers for change are globalization — the elimination or reduction of trade barriers — and deregulation of industries, picking up speed and making themselves felt strongly during the last decades.

The opening up of formerly closed, monopolized, or heavily regulated markets like airlines, telecommunications, energy supply, railway services, banking, etc., has created sizable space for innovation organizations like Ryanair, Virgin, Enron (though probably not the best case to talk about today), Vodafone, and *even* the financial crisis (a case of innovation results worse than Enron's, however). All the companies involved are of tremendous interest from an innovation management point of view — but only partly can their development be explained by technology.

Applying terminology from the field of economics: this stresses the importance of noticing the ways in which "institutions" change and how such changes affect the opportunities for innovation. The possibility of various legal actions is always of importance from an innovation point of view.

We might tend to forget it, but, often, more industrial phenomena than might have been expected are connected to institutional changes. One could, for instance, easily overlook the fact that one of the main reasons for the American dominance in management consulting (read: McKinsey, Accenture, Boston Consulting Group, etc.) has a clear connection to the 1930s legislation preventing US banks from performing audits.[13] There is more of design embedded in these companies than one might assume at first.

The mere existence of the European Union, and its many corollary mandates, as well as the political handling, or ignoring, of something as pervasive as climate change issues are recent, and most likely also future, examples of pertinent high-impact influences.

It is really thrilling to read the monumental work *"The Commanding Heights,"* by former Economist journalists Daniel Yergin and Joseph Stanislaw, describing how the opening up of entire countries and how massive economic deregulation changed the world.[14] It turns out that personalities like Margaret Thatcher, Manmohan Sing, Michael Gorbachev, and Deng Xiaoping ought to be of huge interest for innovation management scholars. What they accomplished during their times is just stunning. They created an open arena for the myriad of business innovations that we experience and ponder today. They "managed/designed" opportunities, which, in the longer run, made it possible for others to "manage/design" businesses as well as other novel opportunities like, for example, social ones.

Once John Micklethwait and Adrian Wooldridge suggested that globalization be "A future perfect," aptly recalling the state of the world before the two World Wars.[15] That time was an era of innovation. In some sense, it was a rather chaotic period — but entrepreneurial to boot. Now a similar period seems to be in the offing again.

When pondering innovation management, do not forget the conditions, prerequisites, and settings for it. Coping with innovation often means coping with *turbulence*.

Managing yourself — and the bold innovative ones around you

Instead of focusing on external factors like those discussed so far, emphasis should be placed on internal factors as the foundation and task for change management — thus factors such as someone in the organization wanting to change something; more broadly, activities performed *within* a corporation.

So engineers creating new kinds of solutions or marketing people suggesting new kinds of solutions provide good examples of internal factors. Designers, wherever placed or acting, would also serve as examples of internal forces. And please do recall that entrepreneurial individuals do not necessarily have to reside at the top of an organization.

Academics sometimes aggregate the internal factors into the concept of "entrepreneurship." It is probably no happenstance that this is one of the sub-disciplines of management research that has grown most rapidly in recent decades. It is also a field of research that, during the last decades, has developed in parallel to other growing academic fields like "Change management," "Innovation management," "Learning organizations," "Knowledge management", and "Management of Creativity."[16] In practice, all these fields of research are interrelated and there are good reasons for letting them cross-pollinate. In real life, isolated academic disciplines do not exist.

Doing research in these areas, we easily forget the risk takers behind revenue-generating innovations. It must be underlined, even though entrepreneurship may be distributed and shared: people like Sam Walton (Wal-Mart), Richard Branson (Virgin), Ingvar Kamprad (IKEA), Michael O'Leary (Ryanair), Steve Jobs (Apple), Serge Brin and Larry Page (Google) do bring a certain flavor to the field of innovation. They are often considered as unusual, even unique, people — bold, risk taking, but always passionate in their undertakings.

In a quite entertaining handbook, *"Think Like an Entrepreneur: Your Psychological Toolkit for Success,"* the claim is put forward that entrepreneurship is a state of mind — not a job.[17] This has, of course, been stated any number of times before, but, precisely because of this, it is worth contemplating.

In practice, this means: if you attempt to innovate — expect *resistance.*

What will follow presently is a sequence of illuminating marketing cases borrowed from Bob Liodice concerning ten unusual ideas that really changed the world of marketing.[18] Read them, and be inspired. However! Do also notice that they are considered as exceptions — way outside the mainstream of marketing.

Idea No. 1: L'eggs' packaging hatches a new look

In 1969, designer Roger Ferriter' of Herb Lubalin Associates felt deeply unsatisfied. It was in the morning of the day when he would be presenting new marketing and packaging ideas to Hanes for a low-cost pantyhose launch — and he judged that his work was lacking in creativity. What he sought for was a novel way to showcase the product. While squeezing the pantyhose in his fist to see how compact it could be, it struck him that the package could take the form of an egg. Immediately, he also realized that egg rhymes with leg. Adding a French flair, he named the product L'eggs, prepared sketches for the afternoon session, and hatched one of the most successful product launches in history.

Idea No. 2: Absolut Vodka's bottle speaks for the brand

Lars Lindmark, CEO of Sweden's Wines and Spirits, teamed up with Gunnar Broman, a Stockholm advertising executive, to create an export liquor brand. Bowman borrowed the name Absolut Pure Vodka from an inexpensive Swedish vodka and took inspiration from 19th century pharmacy bottles for the unusual, striking package. Ignoring opposition from art directors, liquor executives, and focus groups, Lindmark began shipping. When he associated the bottle design with the Absolut (something) campaign from ad agency TBWA, the bottle itself became its own *marketing* engine. Launched in 1979, Absolut had become the best selling imported vodka in the US by 1985.

Idea No. 3: Woodbury soap dares to use sex appeal

In 1911, males dominated the advertising industry, and the advertising they produced was predominantly product-centric. It took a woman, Helen Lansdowne, who headed the newly formed women's editorial department of J. Walter Thompson, to challenge the norms of the day. She refocused Woodbury's advertising on product users with ads that featured elegant young ladies enjoying the attention of dashing young gentlemen. The campaign she directed, "Skin You Love to Touch," is considered by several advertising historians to be the first modern ad campaign to rely on sex appeal.

Idea No. 4: Apple Computer — one mention, one airing

The '1984' commercial introduced the Macintosh personal computer to the world at prime time. Aired nationally just once, during the third quarter of Super Bowl XVIII in 1984, the Ridley Scott-directed spot by Chiat Day represented the Macintosh as a means of saving humanity from "conformity," hinting at Orwell's '1984'. It mentioned Apple Computer only once. The board of directors had hated the spot upon viewing it when Steve Jobs and John Sculley asked for permission to run it. Steve Wozniak volunteered to fund half the cost for airing the ad personally, but, luckily, there was no need for that, as the board eventually caved in and gave its approval. That one airing has resulted in much praise, encapsulated in, for example, the following quote from Wikipedia: "It is now considered a watershed event and a masterpiece in advertising."[19]

Idea No. 5: Avis tries honesty

The "We're #2. We Try Harder." campaign broke all of the rules. It admitted that car rental Avis was losing money, was short of customers, and was second to Hertz. Test results from ad agency Doyle Dane Bernbach were so poor that no one would allow it to run today. But Avis CEO Robert Townsend believed in the tagline and knew that it expressed Avis' management's serious ambition to be different, effective, and outstanding. Almost 40 years later, it is more than a catchy slogan — it has become the essence of Avis.

Idea No. 6: Burma-shave lines the highways

In the early years of the automobile, Clinton Odell developed a brushless shaving cream. It was a great product with no *marketing* plan until his son, Allan, pitched him an idea in 1925: consecutive signs with simple verses, posted at the side of highways. Odell, not crazy about the idea, gave Allan $200 for a trial near Minneapolis. The signs delighted motorists, sales soared, and the iconic campaign eventually spawned 600 verses on 7,000 signs, many submitted by the public through an annual contest. Within a decade, Burma-Shave became the second most popular shaving cream in the US.

Idea No. 7: Benetton brings people together

In 1965, four brothers in Treviso, Italy, started the now well-known international fashion empire Benetton. In 1984, still advertising only in

Italy and France, they hit upon an idea. Focusing on the global appeal for racial harmony and peace, Benetton launched its "All the Colors of the World" campaign. In 1989, an intense collaboration between Luciano Benetton and photographer Oliviero Toscani produced a bold and startlingly different campaign. Removing the merchandise from its ads, "The United Colors of Benetton" featured symbolic multicultural photographs. Benetton's commitment to ethnic diversity remains the staple of its advertising today.

Idea No. 8: Volkswagen tells the whole truth

What do you do if you are Doyle Dane Bernbach and the small, ugly, foreign car you are promoting is competing with over-the-top, macho, American superhero cars? You advocate the negative truths about your car: it is not big, beautiful, or fast. Then sneak in the positives: it does not consume lots of gas, oil, or tires and does not require a big parking spot or high insurance premiums. By turning negatives into witty positives, DDB created an influential ad campaign that made the VW Beetle the best-selling imported car in the US and proved that it pays to "Think Small!"

Idea No. 9: Burger King's "Subservient chicken" dominates

The brand promise, "Have it your way," took on a whole new meaning in 2004 with Burger King's "Subservient Chicken" campaign. Launching the new TenderCrisp Chicken Sandwich for the company and targeting young adults, VP Brian Gies wanted to launch the new product in an unconventional way. Crispin Porter & Bogusky's solution was to launch an interactive website featuring a chicken that could carry out seemingly any command visitors typed in. The results? A million hits in a single day, 20 million the first week, 396 million the first year, an average of a remarkable six to seven minutes spent on the site by each visitor and a sales increase of 9 percent a week.

Idea No. 10: Nike takes a $35 logo the distance

One of the world's most recognizable logos was derived from truly humble beginnings. In 1971, Carolyn Davidson, a graphic-design student at Portland State University, met University of Oregon track runner and accounting teacher Phil Knight. Phil and his coach, Bill Bowerman, needed a logo for a line of athletic footwear for their newly started company. They named their product Nike, after the Greek goddess of victory. For $2 an hour, they hired Davidson as their designer, and, inspired by the wing of the famous statue of Nike, she created the swoosh. Total invoice: $35.

These ten ideas are, hopefully, inspiring — even though, as initially stated, they are outside mainstream marketing.

But, when considering the management of innovation: do not forget *who* might be behind it (yourself, or someone you are supposed to be managing?).

Managing the *possible*

From a management point of view, all kinds of changes — technological, social, political, demographical, cultural, personal — come associated with opportunities connected with innovation. It is precisely as management guru Richard Norman states: every disruption of an existing convention is fertile ground for innovation and entrepreneurship.[20]

It is, however, from a management point, often (though not always) preferable to be initiating and leading change instead of being led by it — paraphrasing a change management bestselling book by professor John Kotter.[21]

The obvious idea behind such a preference is that management may control innovation instead of being constricted or forced by it. As a consequence, the term *control* often seems implicitly embedded in the term management as such.[22]

When it comes to *practice*, however, control in the management of innovation often turns into something inherently complex. Resistance to change — an expression that evidently cannot be disassociated from innovation — may, in many cases, be profound. And, honestly: not every change is automatically for the good so why should there be no resistance to innovation? Just consider the number of companies earning sizable revenue from stable and well functioning products or services, highly appreciated by customers. Why ever should such smoothly functioning systems be forced to change? Change is nothing good by definition — nor bad, *per se.*

The sum of this reasoning is that it is important to understand the context for innovation if we want to understand innovation itself, why and how innovations occur; they rarely, if ever, evolve in a vacuum. It is the very context that establishes the agenda for what is *possible* in a particular situation.

This is the basic reason why Dean Shepard and Mark Shanley once decided on writing an entire book on the topic of timing.[23] What is plain impossible one day might be possible, even compulsory, the next day, and vice versa.

Managing paradoxes

Management implies handling numerous paradoxes. There are gazillions of possible innovations. Which one(s) to opt for — as if such decisions could be "mandated" from the top? It means that innovation is a recurrent topic at top management meetings universally, lacking, however, any final, all-conclusive answer.

To management, managing innovation is about learning how to live with all its concomitant paradoxes. When to change? What to change? How much to change, and how to do it? And also: why change? No exact metrics apply.

Is this a promising new technology? Or should we bet on another one? Is that design a delightful one? Or should we stick to the old, or to a traditional, one? Is there an existing market for it or could it create a market for us? And what will happen to existing products if this new one is being launched? What does the market really need — as if customers know? Could we manage all necessary contingencies — as if we internally already know what is needed? What could be made possible?

Sometimes, one might wonder why so few of these apparent paradoxes are brought forward explicitly in innovation theory.[24] Could it possibly be that sizable parts of existing innovation theory lack a profound grounding in studies of practice? Or is it maybe because, in practice, these paradoxes are often disguised to those of us trying to study and understand them?

One particularly interesting paradox in practice, in real life, is that people do not always act on their vocally professed preferences for new innovations; they stick with existing solutions even though these are judged insufficient. So what is asked for is not realized because people abstain from acting on what they claim to wish. This phenomenon will be investigated further in the following pages.

However, first of all, we have to admit, and keep in mind, that the mere idea of trying something new and different from established practice — like thinking outside the box — can be profoundly difficult, encountering obstacles which are hard to overcome. Often, this can be the case even though no deliberate or intended resistance has been established. We might not just *see* any point in trying something different. Our capability for imagining might have its limitations. So we ought to train ourselves in the art of "imagining"[25] — or what sometimes within the business community is labeled as "vision" (seeing what could exist in the future but not yet here). Testing things, trying them out, making them tangible is, therefore, often exceedingly useful.

Mistakes made by others are easy to see, particularly afterwards, but we have severe difficulties in recognizing our own mistakes, particularly while making them. Thus, there are numerous good reasons for learning through failure.

In the following, the idea of learning by failure and by trying in practice will be tested for their potential for generating results. The focus will be on what happens, and even more on why events unfold in a particular way, when a specific innovation first enters the market, a novelty to customers. We will gradually be looking for some generic lessons to be learned. So, let us now assume that the innovation-to-be is about to enter a critical stage in the process: it is to be launched.

Please note that with this description neither do I claim the "truthfulness" of the classical mental model where an innovation unfolds like in a process (following a process-based workflow manual which most project dependent companies have established in order to organize their work[26]), nor do I suggest that this mental model is to be investigated further here. I rather take it to be obvious that innovation most often is far more complicated than being described "as a process." The study of success stories clearly demonstrates that innovation is far more complex to manage than only by following a checklist of steps in a workflow scheme.

With this focus, it is just the meeting-point between innovation and marketing that will be of further interest here. Innovation is located at the crossroads between engineering and marketing, waiting for "the possible."

We will rely upon a case story from innovation in practice.

Test piloting the market for innovation

In 2003, I got an offer I simply could not resist. It was the opportunity of becoming a Swedish test pilot for a new telecommunications operator, established as "3 Sweden" — the Swedish wing of the global company "3" (a company which is a major global business today).[27]

Thus, I got the offer of becoming one of the very first customers globally, testing the new services concomitant with 3G-based cell phone services. It really was an interesting offer, particularly since I was supposed to test the service in full-scale operation, just a couple of months before the company were to introduce its services to the Swedish market.[28]

Technology and markets

It should be recalled that the 3G standard-cum-technology promised far higher speed and better capacity than ever before — the comparison made with the existing, and dominant, GSM cellular telephony.[29] For 3G, video calling by cell phone was, for instance, talked up as a potential killer application. The possibility of cellular picture telephony will be a recurrent theme as the story unfolds.

Other types of new, innovative services being discussed at the time included the introduction of GPS positioning, the downloading of documents, and an uninterrupted flow of news, information, Internet access, etc. 3G was supposed to take innovation in mobile telephony operations one huge leap beyond what GSM-based services could ever manage — or would have the capacity — to do, and this was already well understood by the general market. People expected 3G to become something huge; it was the next obvious step for cell phones.

Indeed these services exist today, and are well accepted by the market. It took some years for 3G to take off, but for those remembering the S-curve shaped diffusion of new technology enunciated by professor James M Utterback, this should hold no surprise.[30]

Today, we also experience a mobile telephony utilization that has progressed much further. Currently, we may consider the iPhone phenomenon with its momentous number of applications and its marvelous design as the top service — but it seems a safe bet to say that 'we ain't seen nothing yet': the future has much much more in store.

However, in 2003, the idea of 3G was what young people today would call "top notch" and "cool." 3G really promised innovation, and innovation in several dimensions simultaneously.

Of course, at that time, the world had recently seen the burst of the dot.com-bubble, and that unpleasant memory might cause problems for companies launching 3G-based telephony. An economic downturn was around the corner. But, at the same time, the number of mobile telephony subscribers was in ascendancy, as was the use of Internet services. It seemed entirely reasonable to regard 3G as the next technological step, even though it might take some time for it to take off in earnest (which, as we have seen, was the case).

A basis for learning

I was given the test pilot offer a year after I had participated in another "3 Sweden" project, being involved in establishing a scenario concerning the future of the telecommunications business, run internally at "3 Sweden." Those taking part in the scenario project — consultants, academics, business people, etc. (me at that time working for an NGO think tank, previously being a consultant) — were focusing on future generations of customers. "3 Sweden" were indirectly interested in getting second opinions on their own business scenarios. We, the invitees to the scenario project, however, never got any kind of information about these internal scenarios. This was a very wise move by "3 Sweden" — if you really want a second opinion, do not tell your first opinion to anyone.

Nearly all of us scenario project participants accepted the test pilot offer; thus, we came to constitute a test group of 20 people.

Business context

Elaborating on the background of this story: it is of interest to know that 3, as a global holding company for 3G operators, was founded by Hong Kong-based Hutchinson Whampoa several years earlier.[31] Their intent was to create one of the first global 3G operators, a goal which they have fulfilled today. Sweden, as a country of one of the pioneers in the field of mobile telephony and with a high cell phone market penetration, was chosen as one of 3's first markets (together with the UK, Austria, and Italy).

The Swedish operation was established as a joint venture with the Swedish industrial holding company Investor AB, even though 60 percent of shares where held by Hutchinson.[32] Headquarters for 3 in Europe were located to London but "3 Sweden" got its headquarters in Stockholm.

The chairman of "3 Sweden" was of Swedish issue and most of top management was also Swedish. The CEO for "3 Sweden," Chris Bannister, however, was an American and a former CEO of the American broadband operator NTL. He had a track record of having carried out similar market establishment trajectories before.

With Bannister at the helm, the company had a CEO with 15 years of experience in the mobile telecommunications industry, having started and managed cell phone companies in Europe and Asia. The chairman of "3 Sweden" explained via media the choice of CEO in the following manner: "With Chris Bannister in charge, we get the combination of solid marketing competence and extensive experience of start-ups in mobile services required for Hi3G [the term used then] to become a success."[33]

The "3 Sweden" headquarters were located just a few kilometers from the downtown sales-office shop where I later collected my first 3G telephone, starting off my test pilot period. Ironically that shop, the very first 3 shop in Sweden, was placed at the floor level of the same building where the headquarters of TeliaSonera, the main competitor on the Swedish market, moved in just some months later.

Further basis for learning

The "3 Sweden" scenario project in which I had participated before becoming a test pilot was established just shortly after the company started operating in Sweden. In addition to the scenario project, there was a host of other activities going on. "3 Sweden" was doing everything from establishing the physical infrastructure to recruiting numerous employees, all at the same time — and over just a few years. I would estimate that during this period this was one of the biggest entrepreneurial ventures in Sweden when it comes to setting up an entire (and intentionally large already at the outset) company from scratch.

I do remember us in the scenario project hotly discussing the future of the mobile phone business while at the same time visiting the main ICT building that was erected for handling switching. It was designed to cover switching, billing, etc., for the entire Scandinavian market and the cost must have run in the hundreds of millions of Euros. The building was full of expensive hardware and large numbers of overseas experts were busy setting it all up. Obviously, very few Swedes then had the experience of establishing a full-scale telecommunications operator from scratch so international expertise had to be brought in.

Becoming a test pilot meant checking in at the sales-office shop in downtown Stockholm, not yet opened to the public, together with the other nineteen pilots. We signed in for a 3G phone, obtaining a rather big NEC e606 (fairly heavy because of the power supply). All of us were supposed to use the phone during a couple of months, commenting on the service, suggesting changes, and reporting whatever problems.

It was suggested that we stay in regular contact with the company's call-center that would collate all our comments.

Testing in practice

This was a most interesting experience. I remember discussing all kinds of minor problems that I had encountered, calling personnel at that call-center — since there were no customers yet, they had vast amounts of time for talking and listening. In particular, I do recall how the GPS service used old and obsolete maps for my home suburb. Fairly soon, better maps arrived on my screen — but I also must assume that this did not happen because of me specifically. Several of the services mutated on several occasions over the entire test pilot period.

The months before launching 3 publicly seemed to be managed like a just-in-time project. I still feel convinced that the whole setting up of "3 Sweden" was a performed in a profoundly impressive way.

When we were to receive the phones for the pilot tests, something quite funny happened as we were queuing in the shop. Yes, we were the first twenty "customers" in Sweden to receive a 3G cell phone, but we had nobody to make a video call to — except to the other members of the group standing there in the room. And we all smiled at this.

I was in a hurry so I actually did not come around to place any video call at that time. Instead, I just collected my phone and ran off to a business meeting. And to be honest: until this date, I still have not placed a single video call on my mobile phone. But I do know that several of the other test pilots have.

Even to me, this fact is actually fairly strange — and when following the story further on you really have to agree. How come that someone — me, in this case! — can be so focused on a killer application like video calling, without actually being interested in at least trying it out for myself? It must be stated: customers/users do really behave in strange ways when they are adopting innovations. My only excuse or explanation is that I, normally, would consider myself a late adopter of new technology.

I also remember this specific situation very well since I just a couple of weeks later called the person responsible for future marketing development. I suggested that the first thing that should happen for somebody receiving a 3G phone in his/her hand would be a video call. The CEO of the company himself, I suggested, would preferably make this video call — of course pre-recorded and automatically transmitted.

He might say something like the following: "Hello, my name is Chris. Welcome to us. I hope you will enjoy our services like this new service innovation we have named video calling that I am relying upon at this very moment for addressing you." To me — neither having any in-depth knowledge about possible technological problems (if they existed) nor caring about all the hurry and consequent competition for time that "3 Sweden" personnel experienced at the moment — this appeared to be a rather simple, but useful, idea.

The head of marketing development also saw it as a rather funny, striking, and creative idea — but as far as I could see, such an initiative did not happen in the following months. My guess was that for the moment they were terribly busy within "3 Sweden" so I did not repeat my suggestion.

Serendipity[34]

During my months as a test pilot, my daily work place was located in downtown Stockholm — I was always either just walking between offices or traveling by sub-way.

One day, I noticed a young kid (probably around eighteen) walking around at the railway central station with a black cone on his head. The cone had small white figures written all over it. Actually, the whole cone was covered by percent (%) figures. They said "3%", "2.5%", "4%", etc. I also noticed the sign he was carrying on his back. The sign was black too but in white letters it said: "Ask me if you are interested." Underneath, the message was signed "GE Capital." This was a funny creative campaign, I reflected.

This kid became increasingly interesting in the following days since I stumbled on several similar cone-carrying young persons. During an entire week, Stockholm seemed brimming with young people walking around with these cones "campaigning for low interest loans from GE Capital." I suddenly encountered them everywhere.

By pure curiosity, I one day stopped one of these youngsters and asked him what they were doing. He told me that he received about 25 Euros per day in payment for just walking around in downtown Stockholm with the cone on his head and the sign on his back. In addition, he received coffee-money. He regarded it as a pretty well paid job. Yes, what a deal: getting paid for having coffee with his friends as well as for walking around in the sun during spring! His only restriction pertained to the paper-map he had been given. He was not allowed to choose his walking routes entirely on his own. GE obviously did not want every sign-carrier to end up at the same place so each had been given a "walking map."

Everyone has encountered odd marketing campaigns. You may often smile at them, but you also tend to remember them.

I do not remember why, but I called the head of marketing development at "3 Sweden" and told him about this experience. I suggested them doing something similar — like recruiting a couple of hundred young people, giving them a mobile phone on loan, paying them to travel around on the sub-way just calling each other with video calls for a couple of days. My idea was that every sub-way passenger would notice the existence of 3 and be induced to subscribe to their service. Remember: video calling at that time was "top notch," "really cool" — and nobody had it; nobody had even *seen* it.

The head of marketing development offered the same comment as when I suggested my previous idea for video calling. He judged that it was a funny and creative idea.

This time, however, he told me that the marketing campaign had already been decided upon at the European headquarters in London. It was too late to do anything different than was already in that plan. He did not tell me the exact plan (and I also forgot to ask — though my guess is that he would not have been allowed to share the plan with me).

I continued using my phone for a couple of months, checking out for minor problems and calling the call-center daily.

During this period, I also had the brief occasional encounter with the chairman of "3 Sweden." He was not just "3 Sweden" chair but served as chairman for the NGO where I was employed at the time, so it was quite natural that we bumped into each other.

This was how I once got an opportunity to talk with him about the ideas I had conceived on video calling. He also judged them to be very funny and creative. We laughed together over the story of the GE campaign when I described it to him, but I do not recall us expounding on this as anything more serious than just a joke. I do remember, though, him telling of how complicated and time-compressed the entire establishing of "3 Sweden" was.

A couple of months later, I watched the TV channel Eurosport. It was a soccer match though I do not remember which teams were involved.

I noticed that the company 3 had placed heavy amounts of advertising there. The company really put their stake in the ground; they were ubiquitous. I saw them during the commercials breaks as well as their physical presence on the side of the soccer field. Later, I also met several ads for the company in a number of newspapers.

It is well known that TV advertising is rather (or even very) expensive, particularly on sports channels and prime time. In contrast, paying a couple of hundred youngsters 25 Euros per head to walk around is rather inexpensive.

Testing in practice and learning by failure

Of course, all stories are open to interpretation and since this entire book is hoping to inspire personal discovery, please feel free to interpret this story in several more ways. Apply it *if* needed, and *as* needed.

The marketing person has left the company ("3 Sweden" has now been restructured and for several years a new CEO is in charge — my guess is that the initial one, Chris Bannister, was brought in just to set operations up and get them running, and with this task completed, the business is more "mature"); the company has a new chairman and sales and revenue are satisfactory.[35] I would be very surprised if anyone in this story would be upset, no matter how you apply it. That is: if there were ever anything to be upset about (which I doubt). Events unfold by their own logic — right?

It does, however, appear to be at least three interesting lessons inherent in the story, so let us review them.

One lesson is that "it is easy to miss good opportunities and creative ideas from the vantage point of the top of an organization." Is this, however, really something specific for the top echelons of managers or is it something ubiquitous, affecting us all? And why should it ever be any easier to find good ideas just because you are at the top of an organization? Is it not rather the opposite? What do you actually manage to observe from the top? This is a strong argument for testing and trying out ideas in practice.

The second lesson might be related to "how to create sales and increase buying behavior among customers." Marketing, like most activities, often falls back on a quick fix following traditional, well-established patterns. So let us just invest heavily in advertising if we need to increase sales. Nobody would probably question or criticize such an idea. Well, indeed: why should anyone question advertising? It has existed for decades — so there must be something to it.

The third lesson might be about the need for innovativeness in actual action and concrete operations — if an innovative idea is to have a good chance to succeed. How come situations like the ones described above appear again and again? It ought to be underlined that the story from "3 Sweden" is most probably not at all unusual. So the problem, the question, and the paradox goes back to why we encounter such massive difficulties of being just as creative and innovative as we often talk about ourselves being — or striving to be.

Education and knowledge, logics and management

Purposely trying to provoke, management consultant Klas Mellander has stated that the power of learning is far greater than the power of teaching.[36] He, like many others previously, has always been rather critical to the mere idea of teaching — implying that teachers often forget what it is all about. Mellander's way of addressing the problem is shifting the focus to the term 'learning', instead of 'teaching'.

In many ways, he does have a point. The important question is, of course, what students really *learn*, not what we teachers prefer to teach. As teachers, like many people in other situations, we easily become egocentric, assuming that students will always follow our ways of reasoning and thereby automatically understand. We easily forget what the students' specific points of departure may be. Learning is not by definition synonymous with teaching.

So let us reconsider: if we just regard teaching as a tool for learning, it can be of major utility. Good teaching creates (not: is!) good learning. And good learning could seldom be called a waste of time.

Edward de Bono, who sometimes is even more critical to the traditional school system, has taken the subject a bit further. He claims that the education system, very early on, entirely scrapped topics such as creativity and inventiveness.[37] Its singular focus came to be "logical thinking." Furthermore, he asserts that this is the case not just for natural science but also for the humanities. Oftentimes, it feels easy to agree with him. Schools everywhere are full of ideas about rationality and logical thinking.

All this logical thinking may often enough be of good use (not least in engineering), particularly if we agree upon its starting points and understand why, and for what purpose, it has been developed. But it means that we need to know what it should be used for and when it is applicable — as well as when it is not: its limitations. Sadly, though, such issues are not always included in education, and, often, not even mentioned.

Therefore, the idea of applying natural science as the basic benchmark for what so often is labeled "knowledge" tends to become alluring (influencing ideas on how new knowledge should be developed too), using the words of science philosopher Stephen Toulmin.[38] Eventually, such benchmarking might develop into a hegemony extending far beyond its fundamental limits — an unlucky development to boot.

These, of course, are no arguments neither against logics and rational thinking nor against the application of natural science. The world would be a much more sinister place without them. The argument is, rather, that knowledge comes in so many shapes and flavors and hegemony is in all likelihood not for the good when it comes to learning — and specifically not in the field of management.

Creativity — what?

It is sometimes suggested that creativity can flourish only if relying upon basic knowledge acquired in a particular field. That is taken, explicitly or implicitly, to lead to the assertion that the school system should restrict itself to focusing on basic knowledge components (such as: if you study engineering, keep to mathematics and engineering — creativity is for later). Sometimes, it is also claimed that creativity and innovation should be regarded as depending on endowed talent; born predispositions, not acquired talent — implying that creativity and innovation should be left out of the school system entirely. Instead, trust would be put in life itself (or in genetic endowment) to support such competency (such as: school will hand you the tools needed, but, later, you will learn how to apply them creatively — that is, if you are not already "born creative").

As a consequence, we teachers interested in creativity and innovation easily succumb to training students in mundane skills like "how to structure and organize innovation," "how to register a new company and write a business plan," or, even worse, "how to make catchy and colorful PowerPoint presentations." These are, to be sure, useful skills, but they also risk limiting the possibilities for learning — particularly if they are the only topics taught. This is, unfortunately, no caricature but all too often a sad mirror of teachers' distorted understanding of the precepts for learning in creativity and innovation.

Having said that much, it would be an overwhelming task to add new meaningful comments to the often underlined tacit contradiction in terms when proposing something like "teaching for creativity and innovation." Of course, on one hand, one might claim that creativity and innovation, like many other faculties, may be regarded as relying on endowed talent. Taken to the extreme, everything is talent, is a gift; nothing about innovation and creativity could be taught or learnt.

That extreme does not hold: of course, all of us can learn, at least something, in order to be a little bit more creative and innovative than we were before and usually are.

The question is, therefore, only *how*, and *what* we can teach (and learn) about creativity and innovation.

Getting hold of some useful keywords

Of course, all of us, in some sense, have different learning (as well as management) styles, even though they tend to change over time and depend on the topic we are learning.[39] Reading, talking, contemplating, listening, discussing; words, pictures, movies — all kinds of tools for learning can be useful even though we tend to exploit them differently over time.

But... no matter what we might call learning styles: we tend to develop deeper knowledge, and by that also better knowledge, in areas where we *experience and discover for ourselves*. Instinctively, we prefer something hands-on, experimenting, in order to learn.

Most of us remember a situation better when we have experienced it ourselves — when our own brain (even hands and feet - preferably also ears and, in particular, the nose) has been active.[40] Ultimately, it does not matter so much in what way we were activated — as long as we *were* activated.

It is probably no coincidence that most people, when having bought a new technical device, open the box and want to test and try it immediately. Of course, there are people who read the manual first, before trying the gadget out. I actually have such a friend (at least, he claims to display such a behavior). But people like that are rare compared with the majority of us — wanting to try immediately; resorting to reading the manual only if and when we end up in a situation where it is really needed, when we are stumped. In numerous models of how learning occurs, the psychological term 'motivation' is applied for this innate behavior.[41] Without motivation, learning can become just about impossible.

It is our instinctive tendency to being curious that makes us start right away. Having a goal makes us focused. We are curious to see the device functioning, which creates a good situation for learning. We are motivated.

Therefore, important keywords for learning something are, obviously, the likes of goal, focus, and curiosity.

By the way: would keywords capturing your personal picture of management look any different?

Living the message

A major challenge for much of contemporary teaching is that young students often have very nebulous pictures of their own goals — and teachers all too frequently do not even try to help them out there, to find, define, or develop their goals. Students might want to learn something, in some general sense, and that might be all for the good. But students who have a focus (a profound reason) tend to learn far more quickly and easily, more deeply and better, than students lacking focus. "I want to learn French since my girlfriend is from France and I really would like to marry her next year — spending our honeymoon in Paris" is most probably a much better reason — a *motivating* one — for learning French compared to the idea of learning French simply because it might be of some utilty in a very general way in the future and, besides, "mom and dad told me so."

One thing teachers can help students with is supporting them in developing goals — in other words: create reasons for learning. Often, that means using something "real," even creating acutely demanding situations. In the field of innovation, for one, there are many opportunities for this. They ought to be enthusiastically employed.

Curiosity, however, another important prerequisite for learning, is probably more complicated to induce. There is even the proposition that curiosity is a state of mind that some just have, and others do not. If that were really the case — which I find difficult to fully believe (even though I do acknowledge that some people tend to be inherently very curious and others less) — teaching would really be in trouble. (When is a healthy child entirely lacking curiosity?)

We know from experience that curiosity tends to function like a virus. "If you really have it in you, you can transmit it to me also." Meeting someone who is dedicated to collecting stamps, and who demonstrates his enthusiasm, can make a person normally totally uninterested in stamps curious too (not least, for me in any case, if that person starts off by telling that stamps may be seen as reflecting how globalization has evolved over time). So, curiosity can actually be transmitted. There might not be a standard recipe for evoking it in students, but teachers can certainly support it — or suppress it, even kill it.

Teachers who are not keen on learning more themselves, not even a slightly bit curious about discovering novelties related to their own subject, are not likely to create the contagion needed for others to be smitten by curiosity.

'You must live, you must walk your talk' is an entirely appropriate saying often used in management literature.[42] It is a very poignant statement — but still, how often is it actually practiced?

Facilitating for good learning

It is obvious but still worth saying. Good learning should support — not suppress — the learner's own curiosity. Good learning should make students active — instead of inactive. Good learning should stimulate brains to work as much as possible — instead of making them passive. All such reasoning definitely holds for learning creativity and innovation. In brief: good learning often equals *discovery*!

A tremendously good lecture, often spontaneously ranked by students as an outstanding one, does not, by definition, create good *learning* — and, of course, the same goes for management presentations. Actually, it might even do the opposite. "The teacher really gave us a brilliant show here — did you notice how s/he managed to bring forward outstanding arguments cruising around in the field of theory avoiding all potential theoretical pitfalls?"

Fine lecturing easily turns into something like a good show, without any learning achieved. We experience an excellent performance, like seeing an athlete doing something extraordinary on TV without the viewers in the least interested in learning why or how. Our brains have been passive — but we profoundly enjoyed the show. And the better the show, the less the viewer's own brain might be working. It might sound strange but a distasteful lecture, horrible in terms of "performance," can thus actually be really useful in creating a good learning effect (it is not, however, suggested that horrible performances automatically equate with good learning!).

Looking at learning from the discovery point of view means that creating learning *situations* becomes equally important to creating good lectures, if not *more* essential. If learning were focused on discovery, teachers would shift focus and spend far, far, far more time on creating learning situations than the time spent on creating four-colored overhead transparencies, deep theoretical arguments, and good shows.

When looking through the chapters of this book, it is rather tempting to claim that one of the most important issues for teachers (or managers), interested in supporting creativity and innovation (by evoking learning rather than teaching), would be to engage most resources in developing interesting situations for work. Focusing attention on interesting problems to be solved and questions to discuss with the students should be the mission. As a consequence, teachers should just follow students along that road, learning more themselves while the process of learning unfolds.

When this is achieved, however, but not before: support students with tools needed for the specific situation created — and they will learn how to use them very, very quickly.

'Facilitator' would most often be a more appropriate term than 'teacher'. It is probably a rather useful term substituting (and shedding interesting light on) the term 'manager'.[43]

Entering the *specific*

Edward de Bono, like many others, strongly believes that tools and methods for supporting creativity and innovation exist or may be developed — and to prove this, he has formulated and extensively applied methods and tools like "lateral thinking," "mind maps," and "six thinking hats." He has profoundly demonstrated his point. Since fertile methods and tools have proven useful and powerful, it is reasonable to suggest that students (as well as employees) interested in creativity and innovation should learn them and train in applying them.

But tools cannot be applied in a vacuum. No, they need a context to be of any utility. "If I am supposed to invest in learning them, I also need a context — a situation, a reason, a task where I can apply them."

There are important problems associated with what is sometimes suggested or demanded: an idea, a concept, a recipe for generating a competency for creativity and innovation in the most general sense; a competency that might be applied "wherever" and "whenever." This is just too demanding — what might work splendidly does so only in a specific context — even though there are people who manage to be creative in several areas simultaneously or consecutively.

So an established perception just has to be underlined: it is extremely difficult to be creative in some general common sense — possibly impossible. Do never taunt anyone: "Please could you be a little bit creative here?" It simply will not help — or only rarely, by happenstance. Thus, it is far easier and more realistic to ask for creativity in relation to a specific goal, a topic, a question, or a problem to be solved.

My point of departure, as has been stated above, is the meeting point between technology and markets. The case discussed above related to a situation when a potential innovation was about to be launched, the company "3 Sweden" getting to launch 3G-based mobile telephony. So a *specific* situation was described, not a general one.

That is a situation where creativity tools might have been applied — or, at least, contemplated or suggested. As a specific situation, we may now elaborate on or experiment with it, discussing what is potentially creative as well as what is not, and how it all evolved (or got stymied).

Please note: the argument here does not have anything to do with whether the advertising carried out by "3 Sweden" should have been done or not — the case is highlighted here for what it might yield from learning, discussing, and discovery points of view.

It is also a situation from which we may learn something about why creativity can sometimes be so difficult to achieve or apply, so that we may arrive at important conclusions about how to overcome such difficulties.

In the following sections, we will gradually return to the *specific*. But the return will start off in the situation of teaching a well-educated scholar.

The well-educated scholar

A useful starting point for the understanding of many professions is, at least to a certain extent, an analysis of the education behind it, reflecting it and its practice. Since education aims at developing individuals that later enter — even constitute — professions (at least that is the major intent, drop-outs unaccounted for), education also reflects on major points characterizing the profession. The skills we learn at school become the skills we tend to apply when having entered the intended profession. So, without overstating: what we learn at school deeply affects our future ways of acting.

This, however, is often carried out without anyone stating it clearly for us, and teachers, in particular, rarely demonstrate to us in what ways our points of departure affect our ways of acting. Well-educated people might very well be totally *unaware* of the impetus associated with their own learning and training.

Neither marketing people nor engineers are always aware that with a hammer in their hands, they are constantly searching for nails to hammer upon (to resort to a classical saying). And since they know how to use a hammer, all problems they encounter are likely to be treated "as if" they were nails. In some sense, that could be by conscious intent, but it is most definitely more than a little useful when one is aware of this fact.

On a theoretical level, one might learn what a box is — and also about the idea of thinking outside the box. But, in practice, it is difficult to break out of the box without actually knowing that there is a box at all. In order to know that there even is a box that you have learned about, you must be able to observe it from the outside.

Let us reflect a bit on the meaning of this reasoning, indirectly using the case. Let us return to the field of marketing.

Seeing the toolbox as a whole

Today, students who learn marketing at school are very seldom taught the basic premises that ruled when the tools they acquire were developed. They learn their tools well, particularly advertising — they get skilled at their application — but they do not always learn *why* the tools were developed into what they are, the circumstances, and the situation(s).

As a consequence, students easily take what they learn for granted — "as if" it is a fact (a truth) and the only, all-encompassing way of doing something.

When looking at marketing education, there are some peculiarities worth addressing and highlighting if we really want to acquire and understand the most frequently applied traditional tools of marketing.

The first peculiarity to be aware of is that most marketing education has its background in ideas and concepts diffused globally during the mid-1900s. In other words, the ideas and tools taught are fairly old. Still, for example, most marketing courses recapitulate the famous 4Ps, which is *the* manifest of marketing ideas and tools from time (almost) immemorial (briefly, 4P stands for Product, Price, Place, and Promotion).[44] They are the "parameters" that a marketer may combine in order to create sales.[45] You might also call them the four major types of tools found when opening the toolbox of marketing.

The 4Ps were established during the 1960s — and are today often referenced to the renowned marketing professor Philip Kotler. He is one of the best established marketing gurus in modern times. To encounter a marketing scholar who has not heard of the 4Ps and Philip Kotler is rather unusual.

The second peculiarity is that these ideas and tools, contrary to common perception, were developed considerably earlier — in effect, in the early 1900s. So, the concepts that marketing students learn today are not just old — they are *very* old. But the embodiment of them, the shorthand, the 4Ps, is an overall concept established as late as in the 1960s.

The third peculiarity (and the most interesting one) is that the origins of these ideas and tools evolved when industry was in a shape distinctly different from now.

So, in essence, this implies that most of today's marketing courses still contain numerous ideas from the early 1900s. And we all learn them, and often also teach them, even without knowing their ancient pedigree.

In his well-worth-reading book *"The History of Marketing Thought,"* Robert Bartels demonstrates that the origins of marketing thought developed in parallel to the transformation of an agricultural economy into an industrial one.[46] In the early 20th century, however, it was discovered that demand was made up by more than simple purchasing power, so that companies needed to become more active in order to sell.

It was simply not enough to concentrate on distribution (the then-term for what we now call marketing).

At that time, advertising and salesmanship demonstrated that customer demand might be increased by influences other than just the existence of supply. Thus, the idea of advertising was developed/invented (so: an innovation), and, as an indicator, the concomitant literature on advertising grew rapidly. More than one hundred and thirty books on the subject of advertising were published before 1905.

Since then, the idea of advertising has taken hold globally (often enough for good reasons).

This, however, implies that the basic concepts relied upon for advertising were developed at a moment when the alternative to action was *no* action, "doing nothing" (but delivery)!

Today, however, the scope (as well as the need) for substitute actions is far broader than that!

It goes without saying that other interrelated basic ideas within the field of marketing have developed in a similar way as well, the most obvious one being the idea of "segmenting" customers into anonymous groups, instead of treating them individually. This is something which nowadays, because of technological development, on the one hand has become far easier than in the early 1900s, and on the other hand, because of customer and competitive development, has become far more of a necessity than in the early 1900s.[47]

But, having learned marketing at a school of today, it would sound odd to suggest anything else than investing in advertising when an innovation is being launched.[48]

The moral here is, of course, again, that a well-educated scholar often suggests what s/he has learned in school. But of even greater importance, the moral is that it is often *forbiddingly* difficult to think outside the box of "your basic learning."

Well-educated creative and innovative people

The way we understand a business venture is based on the way we analyze it, and the way we analyze it is, in particular, influenced by our prior education and working experience. By that, our basic education, implicitly, also forms mental models of management. And as Peter Gärdenfors, professor in cognitive science, has demonstrated: our mental models almost automatically *instinctively* shape the way we act.[49]

In conclusion, we may ask: how is it possible for a well-educated scholar to think out of the box if, for instance, advertising is all that our scholar has learnt in school? The presence of the box called advertising is often so inherent that he or she normally would not even *see* it as a box. Advertising is a tool that s/he has been taught to master — and, of course, that leads to it being applied.

We might have raised the question even more starkly like this: how could a well-educated engineer (or any kind of specialist) ever manage to think out of the box delivered with her basic training? Would that not imply demanding the impossible? Is it even a contradiction in terms, an oxymoron?

Therefore, it might seem straightforward to suggest broader curricula for specialists than those currently dominating the classrooms.[50]

Why not, for instance, have engineering students attend lectures in history, philosophy, the humanities, and art? Why not let marketing scholars into mathematics or engineering? Why not embed hands-on-training in various courses and concurrently use that training for further learning (not only for what we sometimes call "skills")? Why not have students work for a while at a shop floor in an assembly line "doing manual work," or at a design office trying out CAD-CAM, or at a call-center trying to communicate with those "horrible customers"? Or why not have them work at McDonald's restaurants forced to meet the challenge of "creating services with scarce resources" hands-on?

It would definitely create experiences for life and add valuable knowledge assets to the regular curricula (and why not have faculty members share such experiences?). All in all, akin experiences might provide very productive starting points for further learning — in marketing, engineering, as well as in other fields.

Experiences from early in life tend to affect an individual's worldview far longer, and more profoundly, than the ones gained later.

The narrower fields of knowledge become in the future, which happens in most areas with ever increasing degrees of specialization, the more reasonable a concomitant increasing need for people who have broad experiences.

Developing people that combine an in-depth specialization with a broad knowledge and competency spectrum really poses a positive challenge.

Quite reasonably, though, what has been described will take time to develop. What we are asking for cannot be achieved without substantial lead-times. We must not forget that instant learning rarely constitutes in-depth learning. In-depth knowledge simply requires time to develop, and there are not many shortcuts on the road to becoming a "master."

Still: people tending to think more easily out of the box most often feature a broad knowledge and experience spectrum — why not support its development early in life, at school, and in curricula?

Thinking out of the box

Now, let us return to our marketing-scholar. What can then he/she do in the situation given (assuming it is too late for the curriculum to be changed)?

There are at least three alternative possibilities to offer for a marketing scholar who wants to suggest something else than advertising.

One option is that the scholar deliberately tries to search for something different — deliberately breaking with the tradition of marketing just because a deviation would be noted as markedly different. "If you suggest advertising, then I suggest doing no advertising as a starting point." "If you suggest more time spent on technology, then I suggest relying upon the technology we already have as a starting point." This can be a useful way of action, and some of the marketing cases introduced earlier exemplify such a choice.

Another option is dispensing wholesale with what marketing teaching has taught. "If you suggest advertising, then I say: what do you mean by that, what is actually advertising about — in this case"? "If you suggest technology, then I say: what do you mean by that, what contribution can technology provide in this case"? This could also be a useful way of proceeding. It is probably no coincidence that many successful marketing people never studied marketing at school (just study the cases presented earlier).

A third option is to find out — for yourself! — an accurate picture on what actually does work, and does not work, in marketing in your specific field at the very moment. Sometimes, advertising *does* work, particularly if you compare it to the option of doing precisely nothing, but, sometimes, it does not. "What does the situation we have encountered actually look like and what might be fruitful ideas to solve the actual problems we encounter?" "What would the situation look like if we *really tried* to see it from another angle"? It may be advertising, but it may be something totally different. It may be technology, but it may not be. It may be about design — or again: it may not be. What could be made *possible*?

If you decide to go for the third option, you easily end up starting a journey of discovery yourself — and it is probably a journey worth taking.

There must be reasons why creative and innovative people sometimes are judged bold and risk taking. The options presently submitted could most likely imply asking for trouble internally. It also suggests that we should be duly impressed by managers allowing freedom for such strange ideas as the ones suggested in this book. In a sense, they are temporarily abdicating from the conventional role of manager. Creativity and innovation might sometimes equate with "letting go," "giving up control."

No matter, however, what options you decide to try now and in the future, this workbook will help you out. It is a volume full of seemingly topsy-turvy and unusual ideas. The book does not, for instance, seem to care about dividing the world into "technology" or "markets" (or even give the reader a clear "definition" on what design is). It is also a tome that offers several breaks with established conventions (for instance, including a concept like beauty as well as questioning traditional ideas about "organizing"). It is a book that in several instances challenges conventional wisdom. And it is also a book full of inspiring cases.

Not least — it is a book suggesting that you try out, on your own, to learn by testing in practice, experimenting, and learning through failure, what is *called* failure but is often the road to success. For anyone interested in developing the knowledge of managing creativity and innovation, it really ought to give an enticing new point of departure.

As proposed already (there is, indeed, a verse in the some twelve centuries-old Norse Edda to that effect): good learning is rarely, if ever, a waste of time — even if you are well-educated already.

The innovation-inspired manager

Now: it should not sound strange, but inspiring, suggesting reading some of the previous sections once more, but substituting the word *manager* for the word *teacher* and substituting the word *employee* for the word *student*. In many ways, management is like teaching/learning/facilitating. Managers interested in creativity and innovation must also *live* the message of creativity and innovation.

Managing people is most often very complicated — to state the obvious. Managing innovation, deliberately suggesting it to be a concept broader than "managing people," is most often also very complicated — to continue the obvious.

You might have tried management in practice and you might have theoretical models in your head, explicit or (most often) implicit ones, on how to succeed in the management of innovation — even having studied several academic books on the topic.[51]

If so, you might claim (and include in your basic mental model of management) (as engineers sometimes tend to) that technology sets the agenda for the market — and this suggestion does have merit. But you can also claim that markets set the agenda for technology, included in your basic mental model of management, as marketers would be wont to do — and, once more, the point has merit.

Engineers happily start out from technology — indirectly looking for ways to apply it. Marketing people, on the other hand, often start out with customers — indirectly looking for what these might be interested in.

Where, then, do designers start — in searching the possible *between* engineering and markets?

If you did the suggested exercise of reading this text once again (actually the whole book), you might have ended up finding teaching — no, actually: facilitating — as the foundation for management. If so: management of innovation might become like facilitating at the crossroads of engineering and marketing. The more you know about these two different "playgrounds," the better you might become as a facilitator of the field in between.

In practice, innovation management often means balancing between what can be wished for (technology) and what is demanded (customers), searching *the possible*. It resembles, it *is* standing at the crossroads of engineering and marketing. An inspiring position.

A profound comprehension of what mental models underlie the actions of people you have to "manage" is useful — even *indispensable* — if you aim at facilitating an innovative and creative management culture. How else could people be supported and inspired to think outside their boxes? How else could innovation be managed?

Endnotes

---

[1]Fagerberg, J, Mowery, D, & Nelson, R (eds)., 2006. *The Oxford Handbook of Innovation*, Oxford University Press, Oxford.
[2]Schumpeter, J, 1942. *Socialism, Capitalism and Democracy*. Harper Perennial, 3rd ed., NY, NY.
[3]Rosenberg, N, 2009. *Studies on Science and the Innovation Process: Selected works by Nathan Rosenberg*, World Scientific Publishing, Singapore.
[4]Burke, P, 1979. Concepts of Continuity and Change, in Burke P (ed)., The new *Cambridge Modern History XIII*, Cambridge University Press.

[5]See, for example, Wilson, R & Gilligan, C, 2004. *Strategic Marketing Management: Planning, Implementation and Control*, Butterworth-Heinemann, 3rd ed.

[6]Walton. S, 1993. *Made in America — My Story*, Bantam.

[7]Lewis, W, 2004. *The Power of Productivity*, Chicago University Press.

[8]This book would seem to rely implicitly on a similar definition.

[9]Vise, D, 2008. *The Google Way*, Pan.

[10]There is a marvelous clip of the speech he gave on that occasion to be found at Youtube, see, for example, http://www.youtube.com/watch?v=UF8uR6Z6KLc.

[11]This shift actually happened many, many decades ago, see, e.g. Maddisson, A, 2001. *The World Economy: A Millennial Perspective*, OECD.

[12]Tidd, J & Hull, F (eds)., 2003. *The Service Innovation: Organizational Responses to Technological Opportunities and Market Imperatives*, Imperial College.

[13]McKenna, C, 2006. *The World's Newest Profession: Management Consulting in the Twentieth Century*, Cambridge University Press.

[14]Yergin, D, & Stanislaw, J, 1999. Commanding Heights, Touchstone.

[15]Micklethwait, J & Wooldridge, A, 2001. *A Future Perfect*, Random House Books, NY, NY.

[16]For some basics, see, for example, Drucker, P, 2007. *Innovation and Entrepreneurship*, Butterworth-Heineman; Senge, P, 2006. *The Fifth Discipline*, Random House or Henry, J 2001 *Creativity and Perception in Management*, Sage.

[17]West, C & Steinhouse, R, 2008. *Think Like an Entrepreneur: Your Psychological Toolkit for Success*, Prentice Hall.

[18]Liodice, B, 2010. A look back at 10 Ideas that changed the marketing world, *Advertising Age*, Vol. 81. 15 February 2010.

[19]http://en.wikipedia.org/wiki/1984_(advertisement), 2010-03-23.

[20]Normann, R, 2001. *Reframing Business: When the Map Changes the Landscape*, John Wiley & Sons.

[21]Kotter, J, 1996. *Leading Change*, Harvard Business Press.

[22]For an example where the mere idea of control is questioned, see Streatfield, S, 2001. *The Paradox of Control in Organizations*, Routledge.

[23]Shepard, D & Shanley, M, 1998. *New Venture Strategy: Timing, Environmental Uncertainty, and Performance*, Sage Publishing.

[24]An interesting exception is Farson, R, 1997. *Management of the Absurd, Paradoxes in Leadership*, Simon & Schuster.

[25]A good tool for this is Silverstone, L, 2009. *Art Therapy Exercises: Inspirational and Practical Ideas to Stimulate the Imagination*, Jessica Kingsley Publishers.

[26]For such literature see, for example, Lewis, J, 1989. *Mastering Project Management: Applying Advanced Concepts of Project Planning, Control, Evaluation and Resource Allocation*, McGraw-Hill Professional.

[27]In 2010, the company had more than 18 million subscribers and operated in more than ten countries. In Sweden, the company generated profit for the first

time ever last year (after a ten-year period of introduction and growth). Some 2.0–2.5 Billion Euro have been invested in the project.

[28]The introduction of 3G in Sweden was carried out by operators that had been attributed licenses by a government agency three years earlier, the operators being two established ones (Vodafone and Tele2/Telia) and one a newcomer, "3 Sweden."

[29]The standard is called UMTS (Universal Mobile Telecommunications System) and is one of the globally accepted 3rd generation mobile telephony standards. Its characteristics display several differences to GSM, the second generation standard — not least in the way users access the network.

[30]See, for instance, Utterback, J, 1996. *Mastering the Dynamics of Innovation*, Harvard Business School Press, 2nd ed.

[31]Hutchison is a holding company committed to innovation and technology. It dates back to the 1800s, has more than 200,000 employees and operates in more than 50 countries. It has five core businesses: ports and related services; property and hotels; retail; energy, infrastructure, investments, and others; and telecommunications.

[32]Investor is a Swedish-based industrial holding company with a long history and a track record of innovation. Investor is, for instance, one of the main owners of the telecom equipment (and now services) supplier Ericsson.

[33]See MyNewsdesk, February 28th 2001.

[34]This if probably one of the best terms *ever* coined concerning the concept of accidentally stumbling on unexpected, but useful, knowledge. It has its background in the tale of the *Three Princes of Serendip* (Sri Lanka) written by Horace Walpole in the 18th century. It is nowadays a concept frequently employed in order to describe and understand scientific discovery, see Roberts, R, 1989. *Serendipity: The Accidental Discoveries in Science*, John Wiley & Sons.

[35]See Veckans Affärer, *Vinstmaskinen*, March 2010, pp. 50–55.

[36]Mellander, K, 1993. *The Power Of Learning: Fostering Employee Growth*, McGraw-Hill.

[37]De Bono, E, 2009. *Think! Before it is too late*, Vermilion.

[38]Toulmin, S, 2003. Return to Reason, Harvard University Press.

[39]For an overview on how wide the range of different theories concerning learning can be, see, for example, Jarvis, P, Holford, J, & Griffin, C, 2003. *The Theory and Practice of Learning*, Routledge.

[40]One should not overlook the fact that several fields of research are still puzzling over questions on how we remember/learn, etc. For an overview of the field, see Cooper, L, 1994. *How We Learn, How We Remember: Toward an Understanding of Brain and Neural Systems*, World Scientific Publishing, Singapore.

[41]See Shell, D, et al., 2009. *The Unified Learning Model: How Motivational, Cognitive and Neurobiological Sciences Inform Best Teaching Practices*, Springer.

[42]This saying has an interesting application in Collins, J, 2001. *Good to Great*, Harper Business. A rather blunt attack on the issue (definitely worth considering) is otherwise Fugere, B, 2005. Hardaway, C and Warshawsky, J, *Why Business People Speak Like Idiots: A Bullfighter's Guide*, Free Press.

[43]In modern "general" management-related literature, the term 'facilitator' is (still), surprisingly, rarely used.

[44]The interested may consult, for example, Kotler, P & Armstrong, G, 2009. *Principles of Marketing: Global Edition*, Pearson Education.

[45]Within the academic field of marketing, this is called "the marketing mix" — a term often credited to Borden, N, 1964. The concept of the marketing mix, *Journal of Advertising Research*.

[46]Bartels, R, 1976. *The History of Marketing Thought*, Grid Pub.

[47]See Anderson, C, 2006. *The Long Tail: Why the Future of Business is Selling Less of More*, Hyperion Books.

[48]The curious reader is recommended to look into the fact that the value of advertising has been hotly debated within the community of marketing particularly during the last decade (though much, much longer), see, for example, Ries, A & Ries, L, 2002. *The Fall of Advertising and the Rise of PR*, HarperCollins or Engseth, H, 2006. *The Fall of PR and the Rise of Advertising*, DetectiveMarketing.

[49]Gärdenfors, P, 2006. *How Homo Became Sapiens*, Oxford University Press, Oxford.

[50]Of course, the idea brought forward here is nothing new, see, for example, Kelly, A, 2004. *The Curriculum, Theory and Practice*, Sage Publication.

[51]If not, the following provides a reasonably up-to-date introduction to the field of innovation management: Tidd, J & Bessant, J, 2009. *Managing Innovation: Integrating Technological, Market and Organizational Change*, John Wiley & Sons.

Chapter 12

ACKNOWLEDGEMENTS

As this book builds upon a previous work by a team of researchers, *"Design-Inspired Innovation"* (DII, for short; published in 2006), I am in great debt to my then co-authors, James M (Jim) Utterback, Eduardo Alvarez, Sten Ekman, Susan Sanderson, Bruce Tether, and Roberto Verganti. In addition, I learnt a lot from Scott Cooper, our editor. As we have for more than thirty years, Jim and I have continued to discuss innovation-related issues frequently (well, other subjects as well...), profoundly reflected in the present work, even if not explicitly attributed to him — this is my opportunity to demonstrate my deep gratitude: thank you greatly, Jim! In 2009, Roberto published his own book within the same broad field, *"Design-Driven Innovation"*, which I have sometimes drawn upon and must recommend warmly: it is eminently readable.[1] Bruce, too, has been much present during my struggle with this Workbook as I have frequently utilized his multifold contributions on design, much of it provided during our joint efforts for DII but also with more recent additions. The bulk of the material I have relied upon has been collated from the web, and it is a first for me that the Internet bests literature sources. So I owe a great debt to *Business Week*,

Fast Company, a host of trend analyst firms whose work I am a subscriber to (they are listed in Chapter 4), and innumerable others. I have chosen to highlight these sources where they are explicitly drawn upon.

In April 1997, I was working on a book (in Swedish) on trends for the future in innovation and creativity, and when I spoke warmly about experiences from California, Jim leant on me to visit the larger Boston area once again. Together, we produced a list of people to interview, and labs and corporations to visit. I had read about interesting contributions to innovation provided by industrial designers, and Jim arranged — he was, himself, in a board meeting so he could not attend — for me to meet with Design Continuum, specifically mentioned in the piece I had read. I am very grateful to them for what was a true epiphany: design-inspired innovation, indeed!

All or most of those 40+ design firms that we visited and inter-viewed for DII provided insights drawn upon for this book again, and the financial support provided by VINNOVA, the Swedish Governmental Agency for Innovation Systems, for the project resulting in DII now receives some further payback. In November 2008, Bengt Olsson defended his thesis at the Mälardalen University.[2] Since I served as his thesis advisor, I was as much at the learning and experiencing end as was he. Two design firms in particular opened their processes to Bengt's detailed perusals; he videotaped discus-sions and brainstormings (so called — cf. Chapter 6), obtaining an enormous amount of material that I have drawn upon particularly in the creativity chapter. So, here, thanks are due to the two consultancies

Reload Design and Ergonomidesign. Together, as part of Bengt's thesis work, we arranged a series of Dialog Seminars following the procedure established by Bo Göranzon.[3] I can only join in Bengt's gratitude to the Dialog participants striving to lay bare, to understand the silent knowledge underlying industrial design — and, as part of our chosen process, music production, improvising: improvisation as an essential component to design thinking.

Presentations that I have made to a variety of forums have generated comments directly as well as after further considerations. Forgoing such presentations given before the publication of DII in late 2006, the following list is not exhaustive but somewhat representative: Forum for Innovation Management (Sweden), University of Girona (thanks to Jaume Valls Pasola who, at the time, was a professor there, later having moved on), "my" division of the Royal Swedish Academy of Engineering Sciences, and informal discussions at the Inventors' College (Uppfinnarkollegiet).

As in the previous book, I take the opportunity to express my gratitude for many an exchange with Hans Rausing; largely, it has not been about design particularly, but over the subject writ large, certainly creative, with constructive criticism, taking the broader view, and widening the scope, going for the fundamental issues. The same applies to discussions with Carl-Göran Hedén, much missed, Bengt Stillström, and Henrik Blomgren; enlightening, the true meaning of our conversations is something for me to discover for this book.

Even more than to Jim, my profound gratitude must be directed to Gull-May, for reading and criticizing; for discussing and providing ideas; for sharply reminding me of my blind spots (well, not all of them); and for bringing her own work as an artist, as well as that of artists-cum-designers like Salvador Dalì and Magritte, to bear upon the content here. And for being her — some phenomena in life we experience, even beyond design.

Endnotes

[1]Verganti, Roberto, 2009. *Design-Driven Innovation.* Harvard Business School Press, Boston, MA.

[2]Olsson, Bengt, 2008. *Beskrivningsspråk i och för kreativ praxis: Idéutveckling under gruppsession.* Doctoral dissertation (in Swedish), Mälardalen University, Eskilstuna.

[3]A Dialog seminar manual may be downloaded from http://www.dialoger.se/sida.asp?rubrik=44.

Chapter 13
ABOUT THE AUTHOR

Bengt-Arne Vedin, PhD, is a professor emeritus, now with the Department of Industrial Economics and Management at the Royal Institute of Technology in Stockholm, Sweden, a school where he has previously been a student, teaching assistant, lecturer, and an adjunct professor. He was a professor at Mälardalen University in Eskilstuna, Sweden, and served as a guest professor in Bangkok and Girona, Spain, and also, for six years, belonged to the board for Halmstad University, Sweden. The author of 70 books on themes like innovation, information technology, and future studies, he has served on the boards of about fifteen corporations and a roughly equal measure of boards for professional societies. He is a Fellow of the Royal Swedish Academy of Engineering and of the World Academy of Art & Science, serving on the board for the Foresight Network and the Editorial Board for FUTUREtakes magazine. He is senior advisor to InnovationManagement.se. He has finished more than thirty marathon races.

INDEX